Springer Theses

Recognizing Outstanding Ph.D. Research

Aims and Scope

The series "Springer Theses" brings together a selection of the very best Ph.D. theses from around the world and across the physical sciences. Nominated and endorsed by two recognized specialists, each published volume has been selected for its scientific excellence and the high impact of its contents for the pertinent field of research. For greater accessibility to non-specialists, the published versions include an extended introduction, as well as a foreword by the student's supervisor explaining the special relevance of the work for the field. As a whole, the series will provide a valuable resource both for newcomers to the research fields described, and for other scientists seeking detailed background information on special questions. Finally, it provides an accredited documentation of the valuable contributions made by today's younger generation of scientists.

Theses are accepted into the series by invited nomination only and must fulfill all of the following criteria

- They must be written in good English.
- The topic should fall within the confines of Chemistry, Physics, Earth Sciences, Engineering and related interdisciplinary fields such as Materials, Nanoscience, Chemical Engineering, Complex Systems and Biophysics.
- The work reported in the thesis must represent a significant scientific advance.
- If the thesis includes previously published material, permission to reproduce this must be gained from the respective copyright holder.
- They must have been examined and passed during the 12 months prior to nomination.
- Each thesis should include a foreword by the supervisor outlining the significance of its content.
- The theses should have a clearly defined structure including an introduction accessible to scientists not expert in that particular field.

More information about this series at http://www.springer.com/series/8790

Theo Elmer

A Novel SOFC Tri-generation System for Building Applications

Doctoral Thesis accepted by
The University of Nottingham, UK

Springer

Author
Dr. Theo Elmer
Department of Architecture and Built Environment
Faculty of Engineering
The University of Nottingham
Nottingham
UK

Supervisor
Prof. Saffa Riffat
Department of Architecture and Built Environment
Faculty of Engineering
The University of Nottingham
Nottingham
UK

ISSN 2190-5053 ISSN 2190-5061 (electronic)
Springer Theses
ISBN 978-3-319-46965-2 ISBN 978-3-319-46966-9 (eBook)
DOI 10.1007/978-3-319-46966-9

Library of Congress Control Number: 2016953637

Printed on acid-free paper

This Springer imprint is published by Springer Nature
The registered company is Springer International Publishing AG
The registered company address is: Gewerbestrasse 11, 6330 Cham, Switzerland

Supervisor's Foreword

In response to the critical need to decarbonise the built environment, alternative methods for more effective energy utilisation need to be explored including tri-generation systems. The thesis presents the design, development and testing of a novel proof-of-concept tri-generation system based on solid oxide fuel cell (SOFC) and liquid desiccant air conditioning technology to provide electricity, heating and cooling to building applications. No previous work has been reported on such a system. The theme of the work sits within the topics of low-carbon and sustainable energy technologies, building services and low-carbon building applications. The research has critically examined, both theoretically and experimentally, the novel tri-generation system concept—with encouraging system performance demonstrated. The thesis establishes the significant potential of the novel tri-generation system in providing effective built environment decarbonisation through decentralised generation, strengthening the case for a future hydrogen economy.

The four-year Ph.D. was funded by the EPSRC CDT in Fuel Cells and Their Fuels. The author also received support from the European Commission under the Fuel Cell and Hydrogen Joint Undertaking Initiative (FCH-JU) for the "Durable low temperature solid oxide fuel cell Tri-generation system for low carbon buildings" project (agreement No. 303454). Numerous international conferences were attended and journal articles published relating to the doctoral work. The author was also successful in securing funding from the EU Marie Curie International Research Staff Exchange Scheme for a research secondment to China.

Denmark
June 2016

Prof. Saffa Riffat

Abstract

In response to the critical need to decarbonise the built environment, alternative methods for more effective energy utilisation need to be explored including tri-generation systems.

Tri-generation is the simultaneous generation of electricity, heating and/or cooling from a single fuel source. Solid oxide fuel cell (SOFC) and liquid desiccant demonstrate many characteristics that make them an attractive option in the development of an efficient and effective tri-generation system. SOFCs have high operational electrical efficiencies and a thermal output in good agreement with the low-temperature regeneration requirement of liquid desiccants.

The aim of this thesis is to design, develop and test an efficient and effective proof of concept tri-generation system based on SOFC and liquid desiccant air conditioning technology for building applications. An extensive review of the literature shows that no previous work has been reported on such a system. The research has critically examined, both theoretically and experimentally, the novel tri-generation system concept.

Simulations show tri-generation system efficiencies of up to 71 % are achievable at a 1.5 kW_e capacity, which are encouraging values for a system of this size. An integration analysis, based on empirical data, provides good agreement with the simulations. At a 1.5 kW_e output, a tri-generation efficiency of 69 % has been demonstrated. The inclusion of liquid desiccant air conditioning provides an efficiency increase of up to 15 % compared to SOFC electrical operation only, demonstrating the merit of the novel tri-generation system in applications that require simultaneous electrical power, heating and dehumidification/cooling.

An experimental system, using a microtubular SOFC, shows the novel system can generate 150 W of electrical power, 443 W of heat or 279 W of cooling. Instantaneous tri-generation system efficiency is low at around 25 %. This is primarily due to the low capacity and poor performance of the microtubular SOFC. Although the performance is low, the experimental results demonstrate regeneration of a potassium formate desiccant solution using the thermal output from the microtubular SOFC in the first of its kind tri-generation system. The thesis has established that a clear operational advantage of the novel SOFC liquid desiccant tri-generation system is the potential for non-synchronous operation. The constant SOFC thermal output can be used to reconcentrate the desiccant solution as a form of thermal energy storage. Unlike thermal storage techniques based on sensible energy, a significant advantage of (chemical) thermal energy storage in the form of strong desiccant solution is that there are minimal losses over time. Using this non-synchronous operating concept, the experimental system can generate an increased peak cooling output of up to 527 W and a daily tri-generation efficiency of 38 %.

An economic assessment demonstrates questionable performance; however, this is anticipated to improve with SOFC capital cost reductions. Environmental assessments establish that emission reductions of up to 51 % compared to a base case system are possible, with the potential for zero-carbon operation with the transition to a pure hydrogen fuel.

The thesis presents the following general conclusions with respect to the novel SOFC liquid desiccant tri-generation system: (1) SOFC and liquid desiccant air conditioning are an effective technological pairing. High tri-generation efficiencies, particularly in hot and humid climates, are demonstrated. (2) Appropriate matching of component capacity is necessary. Overall tri-generation system performance is more influenced by the SOFC component than the liquid desiccant. (3) It is primarily the optimisation of the liquid desiccant component that facilitates effective tri-generation system integration and operation.

The thesis proposes that future work should focus on improving the thermal agreement between the SOFC and liquid desiccant component, accompanied by field trial testing in a building context.

Published Work

Journal Papers

Elmer, T., Worall, M., Wu, S., Riffat, S. 2015. Fuel cell technology for domestic built environment applications: State of-the-art review: Renewable and Sustainable Energy Reviews, vol. 42, pp. 913–931.

Elmer, T., Worall, M., Wu, S. and Riffat, S. 2015. Emission and economic performance assessment of a solid oxide fuel cell micro-combined heat and power system in a domestic building: Applied Thermal Engineering, vol. 90, pp. 1082–1089.

Elmer, T., Worall, M., Wu, S., Riffat, S. 2016. An experimental study of a novel integrated desiccant air conditioning system for building applications: Energy and Buildings, vol. 111, pp. 434–445.

Elmer, T., Worall, M., Wu, S., Riffat, S. 2016. Experimental evaluation of a liquid desiccant air conditioning system for tri-generation / waste heat driven applications: International Journal of Low Carbon technologies

Elmer, T., Worall, M., Wu, S., Riffat, S. 2016. Assessment of a novel solid oxide fuel cell tri-generation system for building applications: Energy Conversion and Management: vol. 124, pp. 29–41.

Book Chapter

Elmer, T. and Riffat, S., 2014. State Of The Art Review: Fuel Cell Technologies in the Domestic Built Environment *In:* Progress in Sustainable Energy Technologies Vol. II, Springer.

Conference Presentation with Paper

Elmer, T., Worall, M., Wu, S. and Riffat, S., 2013. Actual versus ideal performance of a SOFC mCHP unit operating in a domestic building *In*: Hydrogen and Fuel Cell SUPERGEN Researcher Conference (H2FC), 15th–17th Dec 2014, Birmingham, UK.

Elmer, T., Worall, M., Wu, S. and Riffat, S., 2014. Energetic and GHG emission analysis of a residential SOFC micro combined heat and power system *In:* Proceedings of the 13th International Conference on Sustainable Energy technologies (SET-2014), 25th–28th Aug 2014, Geneva, Switzerland.

Worall, M., Elmer, T. and Riffat, S., 2013. Simulation of a desiccant dehumidifier for a low temperature solid oxide fuel cell (LT-SOFC) Trigeneration system *In:* Proceedings of the 12th International Conference on Sustainable Energy technologies (SET-2013), 26th–29th Aug 2013, Hong Kong, China.

Elmer, T., Worall, M. and Riffat, S., 2013. A Novel Solid Oxide Fuel Cell Tri-Generation System for Low Carbon Buildings *In:* Proceedings of the 12th International Conference on Sustainable Energy technologies (SET-2013), 26th–29th Aug 2013, Hong Kong, China.

Elmer, T. and Riffat, S., 2012. State Of The Art Review: Fuel Cell Technologies in The Domestic Built Environment *In:* Proceedings of the 11th International Conference on Sustainable Energy technologies (SET-2012), 3rd–5th Sept 2012, Vancouver, Canada.

Acknowledgements

I would like to express my thanks and gratitude to my main supervisor Prof. Saffa Riffat for securing the funding to allow me to complete this work and for his continued support and valuable advice throughout the thesis. I would like to extend my sincere thanks and appreciation to my supervisor Dr. Mark Worall for his invaluable help, advice, comments and encouragement throughout the research. You have taught me a lot. Sorry for being such a nuisance.

I would like to thank the EPSRC and CDT in Fuel Cells and Their Fuels for their continued financial and academic support. Professor Kevin Kendall and Dr. Shangfeng Du's extensive knowledge and expertise in fuel cell science have been greatly appreciated. I would also like to acknowledge the support from the European Commission under the Fuel Cell and Hydrogen Joint Undertaking Initiative (FCH-JU) for the "Durable low temperature solid oxide fuel cell Tri-generation system for low carbon buildings" project (agreement No. 303454), along with all the partners involved.

Thanks to all the technical staff at the Department of Architecture and Built Environment. Dave Oliver, Dave Taylor, Jonny Moss and Andy Matthews your help, advice and wealth of engineering knowledge have made this work possible.

I am of course grateful to my father, Roger, for his unwavering help and support, particularly with the proofreading. Your desktop screensaver continues to provide me with both amusement and encouragement, even after fifteen years. Thanks to Steph for her love and support throughout, even when I was grumpy. Thanks to Günter and Klaus for making me smile. Finally, thank you to my mother, Vivienne, for always being there and supporting me, in anything I choose to do.

Contents

List of Figures

List of Tables

Nomenclature

Symbols

AA_{TC}	Adjusted annual total costs (£)
$CHKO_2$	Potassium formate (–)
c_p	Specific heat capacity (J kg^{-1} K)
D_{AB}	Mass diffusion coefficient (m^2 s^{-1})
D_h	Hydraulic diameter (m)
dm	Mass transfer (kg s^{-1})
dq	Heat transfer (J)
E_o	Reversible open-circuit voltage (Volts)
E_{OCV}	Non-reversible open-circuit voltage (Volts)
EUAC	Equivalent uniform annual cost (£)
F	Faraday's constant (96,472.44 C mol^{-1}) (C mol^{-1})
$\bar{g}_f$	Gibbs free energy of formation (J mol^{-1})
H	Henry's law constant (wt%)
h	Specific enthalpy (J kg^{-1})
H_c	HMX channel height (m)
h_{fg}	Latent heat of vaporisation (J kg^{-1})
I	Current (Amps)
i	Current density (mA cm^{-2})
I_{cc}	Initial capital cost (£)
i_f	Inflation factor (%)
io_e	Exchange current density (A cm^{-2})
i_r	Interest/discount rate (%)
K	Kelvin (°K)
k^h	Convective heat transfer coefficient (W m^2 K)
k^m	Convective mass transfer coefficient (kg m^{-2} s)
$k_h{}^{all}$	Overall convective heat transfer coefficient (W m^2 K)
$k_m{}^{all}$	Overall convective mass transfer coefficient (kg m^{-2} s)
k^t	Thermal conductivity (W m^{-1} K)

Le	Lewis number (–)
LHV	Lower heating value (J kg^{-1})
LHV_{eV}	Lower heating value in electron volts (eV)
LiCl	Lithium chloride (–)
$\dot{m}$	Mass flow rate (kg s^{-1})
MAR	Moisture addition rate (g s^{-1})
MRR	Moisture removal rate (g s^{-1})
n	Year number
n_{cell}	Number of cells (–)
NPC	Net present cost (£)
P	Partial pressures of gas mixtures (Pa)
p	Pressure (Pa)
Pr	Prandtl number (–)
$\dot{Q}$	Input or output power (J)
R	Universal gas constant (8.314 J mol^{-1} K) (J mol^{-1} K)
Re	Reynolds number (–)
R_{ohm}	Electrical resistance of cell (mΩ cm^{-2})
RH	Relative humidity (%)
R_w	Molar gas constant for water (461.5 J kg^{-1} K) (J kg^{-1} K)
SA_m	Membrane surface area (m^2)
Sc	Schmidt number (–)
SPBP	Simple payback period (years)
T	Temperature (°C)
u	Velocity (m s^{-1})
u_f	Fuel utilisation factor (%)
V	Voltage (V)
v	Volumetric flow (L min^{-1})
$\dot{W}$	Work (J)
w_c	HMX channel width (m)
X	Solution mass concentration (–)
z	Number of electrons (–)

Greek Letters

α	Thermal diffusivity (m^2 s^{-1})
ε_{PX}	Plate heat exchanger effectiveness (%)
η	Efficiency (%)
η_h	Enthalpy effectiveness (%)
η_L	Latent (dehumidifier) effectiveness (%)
λ	Mass ratio of air to fuel (–)
μ	Dynamic viscosity (Pa s)
ω	Absolute humidity (kg_{vapour}/kg_{dryair})
ρ	Density (kg m^{-3})

Abbreviations

CFCL	Ceramic Fuel Cells Ltd.
CHP	Combined heat and power
COP	Coefficient of performance
D/C	Dehumidifier core
E/C	Evaporative core
FiT	Feed-in-tariff
GHG	Greenhouse gases
HMX	Heat and mass exchanger
HP	Heat-to-power ratio
HVAC	Heating, ventilation and air conditioning
ICE	Internal combustion engine
IDCS	Integrated desiccant air conditioning system
IEC	Indirect evaporative cooler
LSM	Lanthanum strontium manganite
NG	Natural gas
Ni-YSZ	Nickel–yttria-stabilised zirconia
PED	Primary energy demand
PEMFC	Proton exchange membrane fuel cell
PX	Plate heat exchanger
R/C	Regenerator core
S/C	Steam-to-carbon ratio
SDCS	Separate desiccant air conditioning system
SE	Stirling engine
SOFC	Solid oxide fuel cell
VAS	Vapour absorption system
VCS	Vapour compression system
WHR	Waste heat recovery
YSZ	Yttria-stabalised zirconia

Subscripts

a	Air
AC	Alternating current
act	Activation
adj	Adjusted
atm	Atmospheric
aux	Auxiliary
b	Bulk
BoP	Balance of plant
DC	Direct current
deh	Dehumidifier

des	Desiccant system
el	Electrical
elec	Electricity
eq	Equivalent
ohm	Ohmic
reg	Regenerator
sol	Solution
th	Thermal
trans	Transmission
tri	Tri-generation
v	Vapour
w	Water
wb	Wet bulb

Chapter 1
Introduction

1.1 Introduction

Humanity is now at a global tipping point. Serious decisions regarding the future of world energy supply and demand need to be made. We face the challenge of addressing the issues of reducing reliance on dwindling reserves of fossil fuels and rising energy demand. Predictions regarding the finite supply of heavily relied upon fossil fuels are bleak. It has been calculated that total consumption of easily exploitable oil reserves will take place by 2050 (BP 2012), natural gas in 70–100 years and coal in the next few centuries (Marbán and Valdés-Solís 2007). Exploration will then have to move to places such as the Amazon and the Arctic, or in the deep ocean, therefore making it more expensive and technically challenging. Furthermore, global energy demand is set to rise. Currently, countries such as China, India and Brazil have large and ever increasing populations. Presently, energy consumption per capita in these rapidly expanding economies is well below that of more economically developed economies such as the UK or North America. However, with the continued economic development of these nations, their energy consumption per capita is predicted to grow to, or even surpass that, of countries such as the UK. In the current global situation where the majority of energy supply is from finite fossil fuel reserves, this poses a grave problem; supply is falling and demand is increasing. Furthermore, the principal method of global energy conversion is through the combustion of these fossil fuels. This process liberates significant quantities of greenhouse gases (GHG) into the atmosphere. It is now of critical importance that GHG emissions associated with energy conversion are substantially reduced in order to limit the effects of climate change and environmental pollution. Agreements such as the 1997 Kyoto Protocol have been established in order to try and mitigate the effects of climate change by reducing the quantities of GHG released into the atmosphere. More recently the UK set out

T. Elmer, *A Novel SOFC Tri-generation System for Building Applications*,
Springer Theses, DOI 10.1007/978-3-319-46966-9_1

in its 2007 Energy White Paper that it would commit to an 80 % GHG emission reduction compared to 1990 levels by 2050 (DECC 2008). The European Union (EU) has committed to reducing CO_2 emissions by 20 % by 2020 compared to 1990 levels (Böhringer et al. 2009). Both the UK and EU targets are ambitious; however there is now a common trend amongst many nations towards aspirations of a low carbon future.

Due to the increasing concern over the provision of future energy supply and climate change, there is a significant interest in the development of alternative clean energy sources and efficient power generators. Buildings consume 40 % of the world's primary energy for cooling, heat and power (DECC 2011). Most of this energy is from electricity generated at centralised power stations; where at present up to 70 % of available energy is wasted. The overall system efficiency is low at 30–40 %, leading to a high waste of energy resources, resulting in considerable CO_2 emissions and unnecessarily high running costs. Reducing the energy consumption of buildings can make a substantial contribution towards attaining the EU's 2020 and the UK's 2050 carbon emission targets. But this will only be achieved by moving from conventional centralised power generation systems to onsite highly-efficient clean micro-generation technology (Jradi and Riffat 2014d; Ellamla et al. 2015; Elmer et al. 2015).

One of the most promising possibilities for clean micro-generation is solid oxide fuel cell (SOFC) technology, which can generate electricity directly through an electrochemical reaction which brings together hydrogen and oxygen. The only by-products are waste heat, water vapour, and depending on the fuel used a modest amount of CO_2. Chemical to electrical energy conversion efficiencies can be over 50 % compared to 30–40 % in combustion processes, such as internal combustion engines and gas turbines. Technical assessments have demonstrated that if combined heat and power (CHP) technology is used with SOFC, the total system efficiency can be as high as 90 % (Berger 2015).

Liquid desiccant systems are used in heating, ventilation, and air conditioning (HVAC) applications where simultaneous maintenance of temperature and humidity control is an important benefit to the user. This technology is often used in tri-generation system applications, where the desiccant system is driven by the heat by-product. If the waste heat from SOFCs is used to drive the liquid desiccant unit, then a tri-generation system will result, supplying not only the power and heat as the conventional CHP technology to the building, but also cooling and humidity control. It has been demonstrated in the literature that the inclusion of liquid desiccant in a tri-generation system configuration can provide significant improvement to total system efficiency (Liu 2004; Jradi and Riffat 2014b).

The aim of this thesis is to design, develop and test an efficient and effective proof of concept tri-generation system based on SOFC and liquid desiccant air conditioning technology. The proposed system will be the first of its kind, and can be used for a range of building applications (e.g. domestic, schools, offices), providing power, heating and dehumidification/cooling to the occupied space.

This section has provided some contextual background and the research aim. This chapter is structured in three main parts:

(1) Section 1.2 develops the research aim, the current research gap, contribution to knowledge and research methodology.
(2) Sections 1.3 and 1.4 provide a brief introduction to the tri-generation concept and the hydrogen economy respectively. These key themes have been included in the introduction chapters as it provides a strong basis to introduce the research rationale, aims and objectives. A more in-depth review of the literature is presented in Chap. 2.
(3) Section 1.5 presents a summary of the thesis aim and objectives, along with the thesis structure.

1.2 Research Overview

The thesis establishes that the built environment requires highly efficient multi-functional generators that can provide energy, directly to the user that is safe, secure and economically viable. The fulfilment of these requirements is based upon the need to address the highlighted issues of reducing reliance on dwindling reserves of fossil fuels and rising energy demand.

The aim of this thesis is the design, development and testing of an efficient and effective proof of concept tri-generation system based on SOFC and liquid desiccant air conditioning technology. The system will supply heating, cooling and electricity to building applications. Compared to a conventional separate generation scenario, energy conversion in a tri-generation configuration offers the potential for high energy efficiency and thus greater energy utilisation, providing a range of technical, environmental and economic benefits. The proposed tri-generation system is truly multi-functional, designed to provide the energy needs of a building throughout all seasons. The concept is illustrated in Fig. 1.1. The SOFC is the core of the system, generating electricity and heat from natural gas, with a

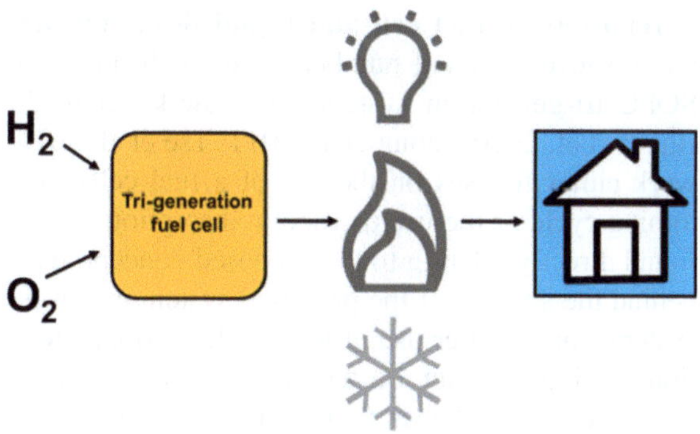

Fig. 1.1 The novel tri-generation system concept

modest amount of carbon generated (hydrogen is extracted from the natural gas, and oxygen from the air). The electrical output from the SOFC is used directly in the building, and if part of a smart grid then it can be imported or exported as required. The SOFC's thermal output will be used for a building's hot water applications and/or to drive the liquid desiccant air conditioning system to produce dehumidified cool air in summer. The cooling output in liquid desiccant based air conditioning systems is achieved primarily through the dehumidification of supply air with some temperature reduction. The liquid desiccant air conditioning cycle is predominately driven by heat for regeneration (some electrical requirement for fans and pumps); therefore it is an attractive option for integration alongside the SOFC.

The proposed tri-generation system is expected to be highly efficient due to the effective simultaneous use of heat and power, providing safe, secure energy directly to the user. The system provides all energy requirements from a single fuel source, thus increasing energy security, and reducing reliance on the national electricity grid. Furthermore, the system could be applied in autonomous applications where connection to any national electrical infrastructure is not possible.

The SOFC and liquid desiccant air conditioning technology will be combined to form the complete tri-generation system. In this thesis, SOFC technology has been used due to its useful thermal output and its ability to internally reform a hydrocarbon fuel, thus reducing the size and complexity of the final system. In the liquid desiccant air conditioning system, in order to overcome the common problem of liquid desiccant entrainment in the supply airstream, a semi-permeable micro porous membrane based cross flow contactor has been employed, operating with a low cost, environmentally friendly, non-corrosive potassium formate ($CHKO_2$) desiccant solution. The work in this thesis has focused on the development of a system suitable for domestic building applications. However, the technological foundations, SOFC and liquid desiccant, are scalable, therefore the developed system could be used for a range of building applications that require simultaneous electricity, heating and dehumidification/cooling.

Extensive literature searches have shown that a SOFC tri-generation system utilising liquid desiccant technology will be the first of its kind: **no research activity is reported on fuel cell and liquid desiccant air conditioning systems**. There are, however, a small number of research publications and patents that focus on SOFC tri-generation systems (Pilatowsky et al. 2007; Al-Sulaiman et al. 2010; Bhatti et al. 2010; Jeong et al. 2011; Tse et al. 2011; Wang et al. 2011). The listed work either focuses on the use of a fuel cells thermal output in a Rankine bottoming cycle or the use of vapour absorption cooling. Although no work has been found directly relating to the proposed concept, the author's rationale and thinking behind the success of the proposed system is that liquid desiccant air conditioning systems make better use of low grade thermal energy compared to vapour absorption cooling systems. Furthermore, liquid desiccant air conditioning regeneration temperatures are lower than that of solid desiccant media (Gandhidasan 2004).

Therefore liquid desiccant air conditioning is deemed a well suited technological partnership for a SOFC combined heat and power (CHP) system, with a temperature output of 50–80 °C (Jradi and Riffat 2014d).

1.2.1 Research Methodology

The methodology of this research is a proof of concept study, utilising three approaches as demonstrated in Fig. 1.2 (1) an extensive review of the literature in Chap. 2, (2) theoretical assessment in Chaps. 3 and 4 and experimental assessment in a laboratory context in Chaps. 5, 6 and 7. The thesis initially aimed to experimentally evaluate the novel tri-generation system concept in field trials at The University of Nottingham's Creative Energy Homes. However, technical and commercial issues with the building-installed SOFC CHP system meant an experimental integration could not be completed. A laboratory based experimental tri-generation system and a theoretical integration analysis, based on empirical SOFC CHP field trial and desiccant system data is presented.

As previously highlighted, no work has been reported in the literature on an SOFC liquid desiccant tri-generation system. Therefore, theoretical and laboratory evaluation of the novel system will provide a clear contribution to new knowledge. Future work should focus on the development of the thesis outcomes in a real building context.

Section 1.2 has provided a brief overview of the research aim, the research gap that exists regarding an SOFC liquid desiccant tri-generation system, the rationale for carrying out the work, and the methodology and structure of the work conducted. Next, the concept of tri-generation is discussed.

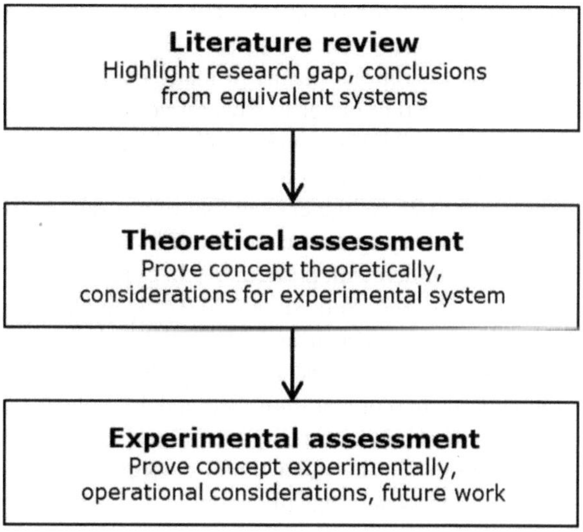

Fig. 1.2 Research methodology adopted in the thesis

1.3 Tri-generation

In recent years, the dramatic increase in fossil fuel prices and the accompanying concerns regarding their environmental impact has driven governments, business and consumers towards cleaner energy resources and to use alternative methods for more efficient energy utilisation.

Combined heat and power (CHP) is defined as the generation of heat and power from a single fuel source, with the view to using both products. CHP is a proven technology, providing various technical, economic and environmental benefits. However, the efficiency of CHP systems tends to fall in hot and humid climates with long summer seasons, where the demand for cooling and/or dehumidification outweighs heating. Tri-generation has been introduced to take the concept of CHP to the next level. By incorporating a heat driven cooling cycle into the system, higher levels of overall efficiencies are achievable due to a greater utilisation of waste heat throughout the year. Energy conversion in a tri-generation configuration offers the potential for high energy efficiency and thus greater energy utilisation (Jradi and Riffat 2014d). Typical cooling technologies in such systems include absorption, adsorption, ejector or liquid/solid desiccant (Wu and Wang 2006).

In small scale (<15 kW_e) tri-generation systems it has been estimated that over 80 % of fuel energy is converted to usable energy, thereby increasing the cost and emission saving potential of adopting a CHP system (Sonar et al. 2014). In many domestic building applications the demand for cooling coincides with a reduction in heating demand, enabling tri-generation systems to operate for a greater period of time over the course of a year compared to conventional CHP systems. Tri-generation systems therefore have the potential to increase access to the benefits achievable from on-site electrical generation, primarily; reduced emissions and operating costs (Sonar et al. 2014). The application of tri-generation systems in the domestic built environment has been increasing, particularly in areas where the thermal demand during the cold season is balanced by an almost equal cooling demand during the hot season (Badami and Portoraro 2009). Recent changes in climatic conditions means in many countries the demand for cooling in buildings is projected to increase. Several northern European cities such as London now report having higher summer cooling loads than winter heating loads. Kolokotroni et al. (2007) state that as a result of the urban heat island effect, cooling loads in cities is up to 25 % higher than that of a rural location over the course of one year. Tri-generation systems are therefore a clear technological option to help address this change in demand, particularly in cities.

Fuel cells are well suited to tri-generation built environment applications because they produce heat when generating electricity, have high electrical efficiency and excellent load-following characteristics (Deng et al. 2011). Moreover, continued technological improvements to fuel cells have, in recent years, increased interest in fuel cell based tri-generation systems (Yu et al. 2010). Due to the advantages of both fuel cell and liquid desiccant technology, a SOFC liquid desiccant tri-generation system has the potential to generate significant energy savings in the built environment, leading to reduced atmospheric pollution and operating costs.

Section 1.3 has provided an introduction to tri-generation systems. Next, Sect. 1.4 discusses the concept of a hydrogen economy, in which the developed SOFC tri-generation system will sit. Particular reference to the drivers and barriers of such an energy system are explored.

1.4 The Hydrogen Economy: Challenges and Opportunities

A future energy system based around the use of hydrogen—the so called 'hydrogen economy' was first coined by John Bockris during a talk given at General Motors in the 1970s (NHA 2005). A shift towards an energy economy centred on the use of hydrogen has been suggested by many as a viable means of moving away from the current intensive use of hydrocarbon fuels, and towards cleaner more efficient means of energy conversion (Berger 2015). Hydrogen may be used as a fuel for motive power (cars and boats), the energy needs of buildings and portable electronics. In such a system hydrogen is not a primary energy source, but an energy vector. This is because hydrogen does not occur naturally in useful quantities and needs to be produced from other sources i.e. water.

Hydrogen, acting as an energy vector, represents significant potential in permitting the extensive use of renewable energy resources (EPSRC 2015). Currently, the main disadvantage regarding renewables is their intermittency. Hydrogen can be used as a way of balancing mismatch in supply and demand. In periods of renewable energy supply, hydrogen can be generated through the electrolysis of water, stored, and then fed through a fuel cell during periods of no/low renewable energy supply. Depending on storage technique, hydrogen may be stored for long periods of time i.e. inter-seasonally, without the issues of discharge over time and cycling limits found in batteries. Although round trip efficiency in such a system will be low, it will provide an uninterrupted supply of clean renewable energy.

Hydrogen as an energy vector exhibits both positive and negative attributes. The main advantage of hydrogen is the absence of harmful GHG when used in a fuel cell to produce electricity. Disadvantages include issues regarding its production and storage. The use of hydrogen will only ever be as clean as its method of production. Currently 98 % of annual hydrogen production (0.1 Gton) comes from the reforming of steam, which is an energy intensive process that involves the combustion of fossil fuel, and cannot be considered a sustainable option for the future. A more promising option for future mass hydrogen production is the electrolysis of water. However, its environmental feasibility is based upon the source of electricity, and currently it is cheaper to generate electricity from the combustion of fossil fuels and not renewables. Furthermore, due to its low volumetric energy density (0.03 kg L^{-1} at 350 bar) compared to other fuels such as gasoline (0.7 kg L^{-1}) (Marbán and Valdés-Solís 2007), the storage and transportation of hydrogen is problematic. As a result, hydrogen storage options are often bulky, heavy and incur large energy penalties in order to store or release the hydrogen. Hydrogen storage options are currently one of the largest challenges facing the development of a hydrogen economy.

A shift towards the use of hydrogen as an energy vector will reduce the dependence on fossil fuels and in turn has the potential to address the current issue of supplying low carbon, secure and economically viable energy. Many counties have undertaken large commitments to hydrogen, fuel cells and their applications. Japan, USA, Canada, Germany and more recently the UK all have large research and development programmes centred on solving the issues and challenges surrounding the use of hydrogen in future energy systems.

It is envisaged that the first generation of hydrogen fuel cells for the domestic built environment will run on hydrocarbon fuels such as natural gas. Following this, the second generation of fuel cells will run on pure hydrogen (Gencoglu and Ural 2009). There are two reasons for the first and second generational stages of fuel use. One, the infrastructure to store, transport and deliver natural gas to individual homes is already present in both the UK and most developed countries. Two, as highlighted above, currently the production of hydrogen is very expensive and, depending on the production method, can produce more emissions than that offset by the use of pure hydrogen. Braun et al. (2006) ran a SOFC CHP system in a residential application under five different system configurations. When the system was fuelled with hydrogen it did not offer any efficiency advantages over methane. It is proposed that the current use of hydrocarbon-fed fuel cells is an essential stepping stone in developing and refining the necessary technology for when the wide-spread transition to clean zero carbon hydrogen production can occur.

The work in this thesis plays a significant role in strengthening the development of a future hydrogen economy, particularly with regards to the efficient use of hydrogen energy in building applications. The SOFC tri-generation system is a highly efficient, multi-functional on-site energy conversion system that can be used to replace traditional heating, cooling and electrical generators in buildings.

Next, the research aim is summarised and the specific objectives of the thesis are described.

1.5 Summary of Thesis Aim and Objectives

The central aim of this thesis is to design, develop and test an efficient and effective proof of concept tri-generation system based on SOFC and liquid desiccant air conditioning technology for building applications. The integration of SOFC and liquid desiccant air conditioning technology will be the first of its kind system, and serves to progress the field of low carbon sustainable energy conversion techniques for building applications.

Seven thesis objectives have been established with the purpose of achieving the thesis aim, and are summarised as follows:

(1) Highlight the current gap in the literature surrounding SOFC liquid desiccant tri-generation systems, particularly for building applications.
(2) Justify the selection of an appropriate working fluid for the liquid desiccant air conditioning system.
(3) Numerically evaluate the liquid desiccant air conditioning system, with particular regards to its suitability for integration in an SOFC tri-generation system.
(4) Development of a system model to evaluate the feasibility of integrating SOFC and liquid desiccant air conditioning technology, and to establish how performance varies with operational changes in the system.
(5) Evaluation based on experimental data, of the SOFC and liquid desiccant air conditioning components.
(6) Tri-generation system integration with energetic, economic and environmental evaluation.
(7) Recommendations for future work related to SOFC liquid desiccant tri-generation systems, particularly for building applications.

Originally, before the technical and economic issues were encountered with the building-installed SOFC CHP system; several technical objectives for the final building-installed SOFC tri-generation system were selected. These are detailed in Table 1.1, and are based upon the EU TriSOFC project (Riffat 2012).

Due to it being laboratory based and of a lower electrical capacity the experimental tri-generation system does not meet the technical objectives set out in Table 1.1. However, the simulations presented in Chap. 4 and the theoretical tri-generation system integration analysis, based on empirical data presented in Sect. 7.2, does aim to maintain the technical objectives.

An electrical output of 1.5 kW_e has been selected based on its appropriateness for providing baseline power in domestic building applications; a value used extensively in the domestic fuel cell CHP sector (Ellamla et al. 2015). Furthermore, 1.5 kW_e is within the 2 kW_e maximum limit to be considered for the UK feed-in-tariff (FiT), therefore increasing the incentive to adopt the developed tri-generation system. Depending on thermal utilisation in the developed tri-generation system, combined efficiency could reach 85 %, similar to that of combustion based systems. However, as demonstrated in Sect. 2.4, for the developed system to compete with current combustion based tri-generation systems it needs to have a higher electrical efficiency. At the proposed scale i.e. <2 kW_e, a combustion based

Table 1.1 Summary of technical objectives of the tri-generation system

Technical parameter	Goal
SOFC electrical power	~1.5 kW_e
System efficiency (based on natural gas)	>85 %
Electrical efficiency (LHV)	>45 %

system could not meet the 45 % electrical efficiency target, an SOFC can, and is therefore a technical objective of the work. Steinberger-Wilckens (2013) states, for hydrogen fuel cells to produce a marked effect on the stationary market they need to match and surpass the performance of current CHP/tri-generation systems. A high electrical efficiency target strengthens the case for hydrogen based energy conversion in building applications.

Reduced system efficiency is often experienced by CHP systems operating in hot and humid climates where the demand for dehumidification/cooling outweighs heating. As a result, the first of its kind SOFC liquid desiccant tri-generation system has been proposed. The novel system aims to overcome the issue of reduced thermal energy utilisation in summer months and thus improve system performance. Although the proposed system operates on a hydrocarbon fuel, it is expected to significantly reduce the associated CO_2 emissions compared to an equivalent separate generation system consisting of a boiler, grid electricity and vapour compression air conditioner. The development of the novel tri-generation system serves to provide an essential stepping stone in developing and refining the necessary technology for when the wide-spread transition to clean zero carbon hydrogen production can occur.

1.5.1 Thesis Structure

The thesis is divided into nine chapters, with the purpose of addressing the seven thesis objectives and thesis aim.

Chapter 1 provides the introduction to the thesis, presenting a research overview, the current research gap, a summary of the research aim and objectives and the thesis structure. Chapter 2 provides an in-depth systematic review of the current literature surrounding: fuel cell (in particular SOFC), liquid desiccant air conditioning, and tri-generation systems; focussing on both domestic and general building application work. Conclusions are drawn that are relevant and specific to the proposed tri-generation system. The aim is to highlight the current gap in the literature regarding a SOFC liquid desiccant tri-generation system. Furthermore, where relevant, literature is referenced throughout the thesis.

Chapters 3 and 4 provide validated numerical assessments of the liquid desiccant and tri-generation system respectively. Numerical assessment has been used to: (1) aid the understating of the performance and operating process of such systems, (2) prove the concept and quantify/qualify the benefits of integrating the two technologies, and (3) provide a benchmark of performance for the experimental liquid desiccant and tri-generation systems presented in Chaps. 5, 6 and 7 respectively. Chapter 3 provides a contribution to knowledge through the assimilation of various sources of data to physically characterise the potassium formate solution in the required operating range specifically for liquid desiccant air conditioning applications. This has not been previously published and available in the literature within the public domain. Chapter 4 provides a clear contribution to knowledge through simulations of the novel SOFC liquid desiccant tri-generation system.

Chapters 5 and 6 present an evaluation, based on experimental data, of the liquid desiccant air conditioning components. Detailed component evaluation is provided to facilitate effective tri-generation system integration. Chapter 5 presents a laboratory scale evaluation, based on experimental data, of a novel integrated liquid desiccant air conditioning system (IDCS). No previous work has been found in the literature regarding an IDCS, thus the work presented is a clear contribution to knowledge. However, conclusions drawn from Chap. 5 demonstrate that the novel IDCS is not suitable for tri-generation system integration. As a result, Chap. 6 presents a laboratory scale evaluation, based on experimental data, of a separate liquid desiccant air conditioning system (SDCS). The SDCS demonstrates great potential for development of an efficient and effective SOFC tri-generation system. Both liquid desiccant component evaluation chapters consist of experimental set-up, results and analysis.

Chapter 7 provides an evaluation, based on experimental data, of the novel SOFC liquid desiccant tri-generation system. No literature is reported on such a system, thus the experimental work presented is a clear contribution to knowledge. Due to unforeseen circumstances with the building-installed 1.5 kW_e SOFC, a 250 W micro-tubular SOFC had to be acquired for experimental integration and testing. Chapter 7 presents the results and analysis from this laboratory scale experimental evaluation. The chapter consists of system configuration, experimental set-up, and tri-generation system test results and conclusions. In addition to this, a theoretical integration analysis, based on empirical 1.5 kW_e SOFC CHP field trial and liquid desiccant system data is also presented.

Chapter 8 evaluates the novel tri-generation systems on an economic and environmental basis. The evaluation compares the novel tri-generation system to an equivalent system based on electricity derived from a thermal power plant, heating from a gas boiler and cooling from an electrical driven vapour compression system. Finally, Chap. 9 presents the conclusions from the thesis and recommendations for future work.

The work presented in this thesis aims to address the current gap highlighted in the literature of a SOFC liquid desiccant tri-generation system for building applications. The integration of these two technologies has not been previously investigated and documented.

References

Al-Sulaiman, F.A., I. Dincer, and F. Hamdullahpur. 2010. Energy analysis of a trigeneration plant based on solid oxide fuel cell and organic Rankine cycle. *International Journal of Hydrogen Energy* 35(10): 5104–5113.

Badami, M., and A. Portoraro. 2009. Performance analysis of an innovative small-scale trigeneration plant with liquid desiccant cooling system. *Energy and Buildings* 41(11): 1195–1204.

Berger, R. 2015. *Advancing Europe's energy systems: Stationary fuel cells in distributed generation*. F. C. A. H. J. Undertaking. Luxembourg, Publications Office of the European Union 2015.

Bhatti, M. S., J. F. O'Brien, I. Reyzin, M. Grieve and S. M. Kelly. 2010. *Solid oxide fuel cell assisted air conditioning system*. USA.

Böhringer, C., T. F. Rutherford and R. S. J. Tol. 2009. THE EU 20/20/2020 targets: An overview of the EMF22 assessment. *Energy Economics* 31(Suppl 2(0)): S268–S273.

BP. 2012. *BP Statistical Review of World Energy*.

Braun, R.J., S.A. Klein and D.T. Reindl. 2006. Evaluation of system configurations for solid oxide fuel cellbased micro-combined heat and power generators in residential applications. *Journal of Power Sources* 158(2): 1290–1305.

DECC. 2008. *Department of Energy and Climate Change (UK): Climate Change Act 2008*. The Climate Change Act 2008 2013, Electronic article.

DECC. 2011. *The carbon plan: Delivering our low carbon future. D. E. A. C. Change*. London: Crown.

Deng, J., R.Z. Wang, and G.Y. Han. 2011. A review of thermally activated cooling technologies for combined cooling, heating and power systems. *Progress in Energy and Combustion Science* 37(2): 172–203.

Ellamla, H.R., I. Staffell, P. Bujlo, B.G. Pollet, and S. Pasupathi. 2015. Current status of fuel cell based combined heat and power systems for residential sector. *Journal of Power Sources* 293: 312–328.

Elmer, T., M. Worall, S. Wu, and S.B. Riffat. 2015. Fuel cell technology for domestic built environment applications: State of-the-art review. *Renewable and Sustainable Energy Reviews* 42: 913–931.

EPSRC. 2015. *Sustainable energy vectors*. 2014, from https://www.epsrc.ac.uk/research/ourportfolio/themes/energy/subthemes/sustainable/.

Gandhidasan, P. 2004. A simplified model for air dehumidification with liquid desiccant. *Solar Energy* 76(4): 409–416.

Gencoglu, M.T., and Z. Ural. 2009. Design of a PEM fuel cell system for residential application. *International Journal of Hydrogen Energy* 34(12): 5242–5248.

Jeong, J., S. Yamaguchi, K. Saito, and S. Kawai. 2011. Performance analysis of desiccant dehumidification systems driven by low-grade heat source. *International Journal of Refrigeration* 34(4): 928–945.

Jradi, M., and S. Riffat. 2014a. Tri-generation systems: Energy policies, prime movers, cooling technologies, configurations and operation strategies. *Renewable and Sustainable Energy Reviews* 32: 396–415.

Jradi, M., and S. Riffat. 2014b. Experimental investigation of a biomass-fuelled micro-scale tri-generation system with an organic Rankine cycle and liquid desiccant cooling unit. *Energy* 71: 80–93.

Kolokotroni, M., Y. Zhang and R. Watkins 2007. The London Heat Island and building cooling design. *Solar Energy* 81(1): 102–110.

Liu, X.H., K.C. Geng, B.R. Lin, and Y. Jiang. 2004. Combined cogeneration and liquid-desiccant system applied in a demonstration building. *Energy and Buildings* 36(9): 945–953.

Marbán, G., and T. Valdés-Solís. 2007. Towards the hydrogen economy? *International Journal of Hydrogen Energy* 32(12): 1625–1637.

NHA (2005) *The history of hydrogen*. Fact Sheet Series.

Pilatowsky, I., R.J. Romero, C.A. Isaza, S.A. Gamboa, W. Rivera, P.J. Sebastian, and J. Moreira. 2007. Simulation of an air conditioning absorption refrigeration system in a co-generation process combining a proton exchange membrane fuel cell. *International Journal of Hydrogen Energy* 32(15): 3174–3182.

Riffat, S. 2012. *EU project proposal: Durable solid oxide fuel cell tri-generation system for low carbon buildings*.

Sonar, D., S. L. Soni and D. Sharma. 2014. Micro-trigeneration for energy sustainability: Technologies, tools and trends. *Applied Thermal Engineering* 71: 790–796.

Steinberger-Wilckens, R. 2013. Fuel cells and hydrogen joint undertaking: Evaluating the performance of fuel cells in European Energy Supply Grids (FC-EuroGrid). *Fuel Cells and Hydrogen Joint Undertaking*.

Tse, L.K.C., S. Wilkins, N. McGlashan, B. Urban, and R. Martinez-Botas. 2011. Solid oxide fuel cell/gas turbine trigeneration system for marine applications. *Journal of Power Sources* 196(6): 3149–3162.

Wang, Y., Y. Huang, E. Chiremba, A.P. Roskilly, N. Hewitt, Y. Ding, D. Wu, H. Yu, X. Chen, Y. Li, J. Huang, R. Wang, J. Wu, Z. Xia, and C. Tan. 2011. An investigation of a household size trigeneration running with hydrogen. *Applied Energy* 88(6): 2176–2182.

Wu, D.W., and R.Z. Wang. 2006. Combined cooling, heating and power: A review. *Progress in Energy and Combustion Science* 32(5–6): 459–495.

Yu, Z., J. Han, X. Cao, J. Han, X. Cao, W. Chen, and B. Zhang. 2010. Analysis of total energy system based on solid oxide fuel cell for combined cooling and power applications. *International Journal of Hydrogen Energy* 35(7): 2703–2707.

Chapter 2
Review of the Literature

2.1 Introduction

This chapter presents a detailed and comprehensive review of the literature relating to the proposed SOFC liquid desiccant tri-generation system. The review is split into five key topics:

(1) Fuel cell technology in Sect. 2.2
(2) Liquid desiccant air conditioning in Sect. 2.3
(3) Combined heat and power in Sect. 2.4
(4) Tri-generation systems in Sect. 2.5
(5) Tri-generation system assessment methods in Sect. 2.6.

The aim of the review is to (1) highlight the current research gap regarding a SOFC liquid desiccant tri-generation system, (2) demonstrate the feasibility of the development of such a system, and (3) provide a comprehensive background study on the state of the art of the technology used. All literature reviewed has been selected due to its relevance in achieving the central aim of the thesis. Due to the breadth of the thesis topic, a large range of literature sources have been utilised from a variety of disciplines including; fuel cell science, low carbon buildings and HVAC technology. Particular areas the review focusses on are system: design, operation, performance, assessment and energy savings. This thesis focuses on a domestic setting but the tri-generation system is applicable to a range of building types. Thus, the reviewed literature has considered a range of sources from a variety of building types. Section 2.7 provides conclusions to the literature review with specific consideration as to their impact on the development of the proposed SOFC liquid desiccant tri-generation system. A published book chapter and journal article, by the author, on fuel cell technology for domestic built environment applications can be referred to here; Elmer and Riffat (2014), Elmer et al. (2015b).

T. Elmer, *A Novel SOFC Tri-generation System for Building Applications*,
Springer Theses, DOI 10.1007/978-3-319-46966-9_2

2.2 Fuel Cell Technology

This section provides an introduction to fuel cell technology. Fuel cells have recently been identified as a key technological option for improving both building energy efficiency and reducing emissions en-route to a zero carbon built environment (Hawkes et al. 2009; Kazempoor et al. 2011). Fuel cells are not heat engines, and thus their efficiencies are not limited by the Carnot efficiency. By combining hydrogen and oxygen in electrochemical reactions as shown in Fig. 2.1, fuel cells have the potential to produce highly efficient electrical power with little or no emission of environmentally damaging pollutants such as CO_2.

The exothermic nature of the electrochemical reactions makes fuel cells ideal candidates for CHP and tri-generation system applications. Invented in 1839, fuel cell technology is by no means new, however, in the past, fuel cells have struggled to flourish particularly in terms of commercialisation and market application. In the past fuel cells have been used in bespoke projects. NASA used fuel cells in some of their early space shuttles to provide electrical power, heat and water. The reasons for this stalled start are mainly cost but also technological reliability, lack of interest from investors and supporting infrastructure (Gencoglu and Ural 2009). Fuel cells are often categorised by the type of electrolyte. This is determined by the type and purity of the fuel and oxidant used and the operating temperature. There are currently six types of established fuel cells on the market (O'Hayre et al. 2006).

- Proton Exchange Membrane Fuel Cell (PEMFC)
- Alkaline Fuel Cell (AFC)

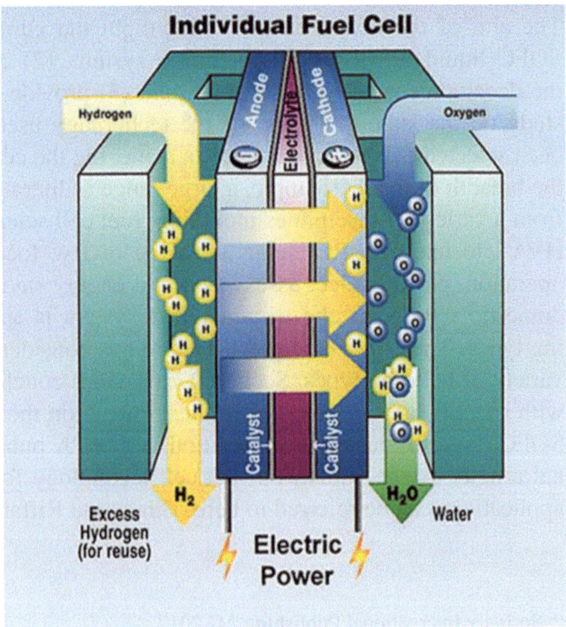

Fig. 2.1 Fuel cell operating concept (ElianEnergy 2011)

- Direct Methanol Fuel Cell (DMFC)
- Phosphoric Acid Fuel Cell (PAFC)
- Molten Carbonate Fuel Cell (MCFC)
- Solid Oxide Fuel Cell (SOFC)

The first three fuel cells are classified as low temperature (80–250 °C), whilst the remaining three are medium to high temperature (250–1000 °C). The operating temperature is often a significant factor when determining which type of fuel cell should be used in a particular application. This is due to a number of factors including; heat usability, start-up time and ability to vary output.

Owing to the variety of types of fuel cells and their modularity, fuel cells have the ability to cover a range of building applications from a single family home to an entire hospital (Kazempoor et al. 2011). Fuel cells are now recognised, across a variety of markets, most significantly the stationary, as a superior technological option compared to conventional combustion based generators (FCT 2013). As a result, the stationary sector is currently the largest user of fuel cell technology, showing year on year growth, demonstrated in Fig. 2.2. In 2012 alone, 125 MW of fuel cells for stationary applications were shipped, a 53 % increase on 2011 figures, representing the rapid expansion of the sector. Furthermore, Ceramic Fuel Cells Ltd. (CFCL) reported that the domestic housing SOFC market is around 17,000 kWe installed per annum, a large market potential. E.ON believes most UK homes are technically suitable for fuel cell CHP, equal to a potential total installed capacity of 24 GWe (Harrison 2012). Fuel cells are an attractive option for building applications because of their high electrical efficiency (even at part load), low emissions, near silent operation, flexibility of fuel use and useful heat output.

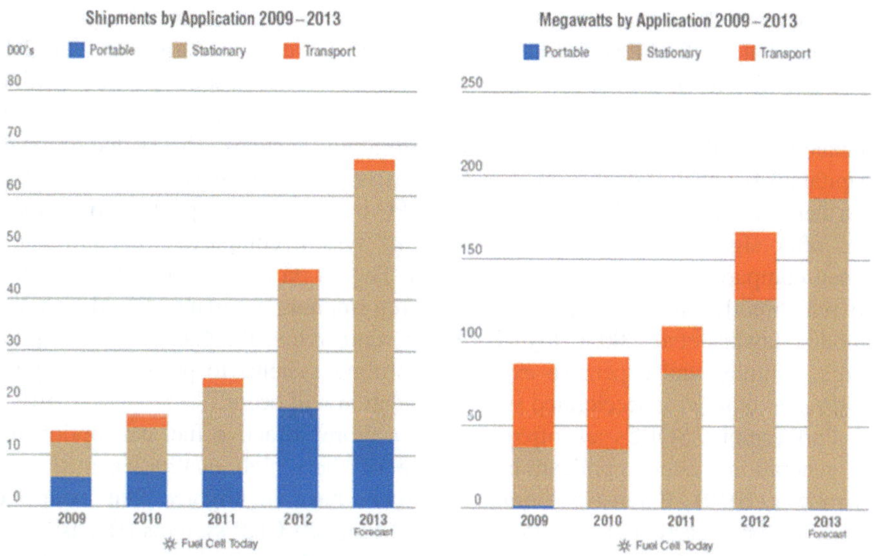

Fig. 2.2 Fuel cell use by application 2009–2013 (FCT 2013)

Table 2.1 Summary of PEMFC and SOFC characteristics (Minh 2004; Wu and Wang 2006; Gencoglu and Ural 2009; Hawkes et al. 2009)

	PEMFC	SOFC
Operating temp. (°C)	30–100	500–1000
Electrical efficiency (%)	35–45	45–60
Electrolyte	Solid polymeric membrane	Solid, stabilised zirconia ceramic matrix with free oxide ions
Charge carrier	H^+ ions	$O^=$ ions
Construction	Plastic, metal or carbon	Ceramic, high temperature metals
Fuels	Hydrocarbons or methanol	Natural gas or propane
Contaminants	CO, Sulphur, NH_3	Sulphur
Cell configurations	Flat plate	Tubular, Flat plate, Planar
Applications	Automotive, stationery	Stationery
Companies	Baxi, Panasonic	CFCL, Ceres
Advantages	Quick start up time, can vary output quickly, compact, no corrosive fluid used	High temperature enables internal reforming, no liquid electrolyte used, useful high temperature heat output can be used in another cycle
Disadvantages	Expensive platinum catalysts required, high purity H_2 required	Long start up time, expensive heat resistant materials needed

Of the six fuel cell variants mentioned, the low temperature PEMFC and the high temperature SOFC demonstrate the greatest promise for early market application, attracting the most attention and investment in building application projects (Sammes and Boersma 2000; Gencoglu and Ural 2009; Kazempoor et al. 2011; Ellamla et al. 2015). PEMFC and SOFC characteristics are summarised in Table 2.1.

Currently it is estimated SOFC developments are around 5 years behind PEMFC, however many commercial developers believe the future of fuel cell technology in the built environment lies with SOFC systems. This is due to lower capital cost as they do not need to use expensive platinum catalysts like PEMFCs. Furthermore, as highlighted in Fig. 2.3, SOFC based systems operating on natural gas offer significant advantages in terms of system complexity and efficiency when compared to PEMFC based systems. This is on account of SOFCs ability to run directly on natural gas, with fuel reformation occurring directly on the anode (Hawkes et al. 2005; Boyd 2008). In comparison, PEMFC systems require large, complex and expensive external reforming systems to produce very pure hydrogen, otherwise accelerated stack degradation will occur.

Furthermore, SOFCs are often cited as a more attractive fuel cell option for tri-generation system applications because of their high quality exhaust heat (Brouwer 2010). Extensive literature searches have highlighted that extensive work has been carried out on domestic scale fuel cell CHP systems (Beaussoleil-Morrison 2008; Beausoleil-Morrison and Lombardi 2009; Farhad et al. 2010; Choudhury et al. 2013). However, only a small amount of work on domestic scale

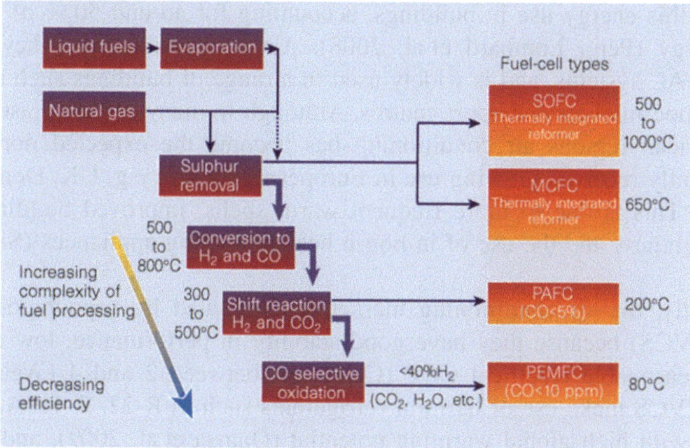

Fig. 2.3 Fuel processing reactions (methane to hydrogen) and their influence on the complexity and efficiency of fuel cell systems (Steele 1999)

SOFC tri-generation system has been found, and this has been predominately simulation based work (Pilatowsky et al. 2007, 2011).

This section has provided an introduction to fuel cell technology and its application in the built environment. Next, Sect. 2.3 presents a review of liquid desiccant air conditioning.

2.3 Liquid Desiccant Air Conditioning

This section presents a summary of the literature surrounding liquid desiccant air conditioning. Section 2.3.1 describes the operating process of a liquid desiccant air conditioning system. The aim is to highlight the significant advantage of adopting liquid desiccant in a SOFC tri-generation system. Section 2.3.2 details the different contactor designs and desiccant working fluids used. The aim is to highlight the benefit of employing a membrane based contactor with a potassium formate working solution. Section 2.3.3 provides a review of different liquid desiccant based air conditioning systems, with particular reference to those operating on waste heat sources. Throughout Sect. 2.3, specific consideration to tri-generation system integration is provided.

2.3.1 Operating Process

As highlighted in Sect. 1.1 buildings consume significant quantities of energy, and thus they are a large contributor to CO_2 emissions. HVAC systems are a major source of this energy use in buildings, accounting for around 50 % of total supplied energy (Pérez-Lombard et al. 2008). Air conditioning is a key function within HVAC systems, and is widely used in a range of buildings such as homes, schools, supermarkets and sport centres. Although in many Middle East, Far East and American regions air conditioning has become the expected norm, it has more recently received growing use in European countries e.g. UK, Denmark and Germany. This is due to more frequent warm spells, improved building insulation/air tightness and the use of in-house heat generating appliances (Smith et al. 2011).

Currently, the air conditioning market is dominated by vapour compression systems (VCS) because they have good stability in performance, low cost, long life and reasonable electrical COP (COP_{el}) of between 2 and 4 (Welch 2008). However, VCS make use of harmful refrigerants such as R-22, R-410A, R-134A, materials with high global warming potential (Ouazia et al. 2009), and consume significant quantities of electrical energy to drive the compressor. Owing to the fact that the most common form of electrical generation in the majority of counties is from the combustion of fossil fuels, VCS can be viewed as neither a sustainable nor efficient air conditioning option (Zhang 2006a). It is thus apparent, with an already high and continually growing global demand for air conditioning there is a need for alternative options that do not rely so heavily on fossil fuel derived electrical energy.

There are a variety of air conditioning systems which reduce the requirement of electrical energy and in place of this use thermal energy to operate. This thermal energy can be sourced from solar or waste heat applications, thus the associated CO_2 emissions from the air conditioning process will be much lower than for the equivalent VCS. This is due to the reduction in electrical requirement and the consumption of waste heat for a useful process. A variety of thermally activated cooling technologies exist, including: vapour absorption, solid adsorption, ejectors and solid/liquid desiccants. Vapour absorption systems (VAS) replace the electrical driven compressor found in VCS with a heat driven absorber and generator, these act in combination as a thermal compressor in the system. Absorption systems have relatively low thermal COP (COP_{th}), in the range of 0.5 in single effect cycles up to 1.2 in double effect cycles (Srikhirin et al. 2001; Welch 2008), which results in the intensive use of thermal energy. Furthermore, due to pressurised operation, the need for high temperature waste heat, expensive and corrosive chemical solutions, e.g., LiCl, LiBr, $CaCl_2$, VAS are relatively large and complex, and this has limited their attraction to many users (Duan 2012) and to applications greater than 10 kW (Pietruschka et al. 2006).

Desiccant air conditioning systems utilise the capability of desiccant materials in removing moisture from an air stream by the natural sorption process. Desiccant systems can be categorised as either solid (adsorption) or liquid (absorption); both types have their advantages and disadvantages. Liquid desiccant systems have lower regeneration temperatures, greater dehumidification capacity and lower air side pressure drop—thus reduced fan power requirement. On the other hand, solid desiccant systems are compact, simple, less subject to desiccant carryover and corrosion. In this thesis liquid desiccant is employed as the thermally activated cooling technology in the tri-generation system. Table 2.2 provides a comparative summary between liquid desiccant and VCS technology.

The information summarised in Table 2.2 illustrates the significant advantages of using liquid desiccant air conditioning, particularly in high humidity applications where waste heat is available. Liquid desiccant air conditioning represents significant advantages compared to both VCS and VAS for efficient air conditioning and tri-generation system applications, these include:

- No harmful refrigerants or working fluids are required
- Reliable and simple, operate at atmospheric pressures
- Can be driven by low grade waste heat (45 °C upwards)
- Decoupling of latent and sensible heat removal. Thermal comfort requirements can be met over a large range of conditions in a low energy, efficient manner
- Effective at cooling capacities of less than 10 kW

A drawback of the use of liquid desiccant air conditioning is that its application and performance is more location/climatic specific compared with VCS and VAS. As demonstrated in the Chap. 3 parametric analyses the dehumidification and regeneration processes are largely influenced by the inlet air conditions.

Liquid desiccant air conditioning systems operate on an open cycle absorption principle and consist of three main components, shown in Fig. 2.4: regenerator, desiccant evaporative cooler and dehumidifier. A supply air evaporative cooler is optional. The system consists of three main flows of fluids, the liquid desiccant solution, water and air.

Figure 2.5a shows a property plot of the desiccant solution process and Fig. 2.5b shows a psychrometric chart of the air process. A simplified description of the operation of a liquid desiccant air conditioning system is provided below.

Outside (supply) air flows through the dehumidifier, coming into contact with strong cool liquid desiccant solution. Because the desiccant solution exhibits a vapour pressure and temperature lower than that of the hot humid air (D1), it

Table 2.2 Comparison between liquid desiccant and VC air conditioning systems (Mei and Dai 2008)

	Vapour compression	Liquid desiccant
Initial investment	Similar	Similar
Operation cost	High	Up to 40 % saving
Driving energy source	Electricity, natural gas	Low-grade heat
Humidity control	Average	Accurate
Temperature control	Accurate	Average
Indoor air quality	Average	Good
Location specific	Slightly	Yes
System instalment	Average	Complicated
Energy storage capacity	Bad	Good

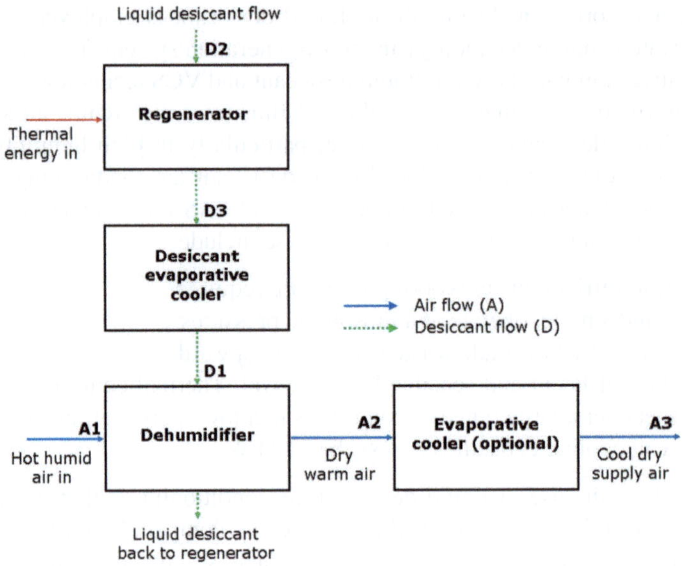

Fig. 2.4 A stand-alone desiccant air conditioning system

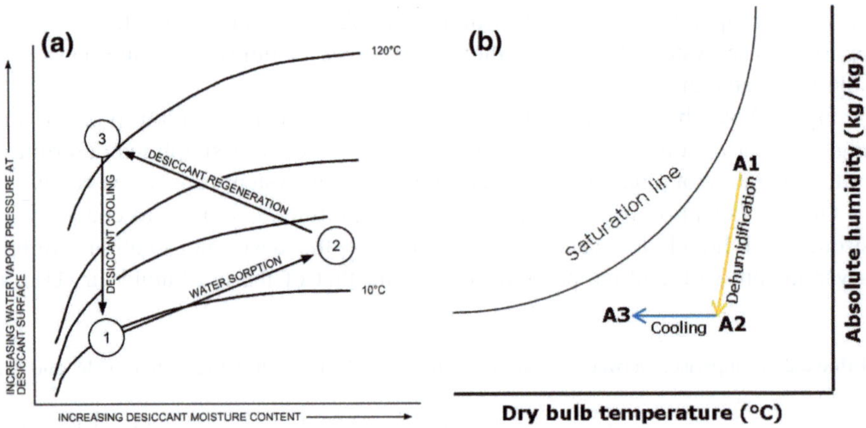

Fig. 2.5 **a** Property plot of desiccant process. **b** Psychrometric chart of air process (ASHRAE 2009)

causes dehumidification and sensible cooling (A1–A2). The supply air can then pass through an optional evaporative cooling device (performs better with drier air) to lower its temperature further (A2–A3). This air is then supplied to the building application.

During the dehumidification process, the desiccant solution becomes weak and warm due to the moisture sorption process (D2). The solution needs to be re-concentrated so that it may be used again. After passing through the dehumidifier, the desiccant solution flows to the regenerator. In the regenerator, the weak and warm desiccant solution has heat applied to it to desorb the moisture added in the dehumidifier; resulting in a strong but hot desiccant solution (D3). In some applications heat may also be applied by heating the regeneration airstream. Following regeneration, the desiccant solution needs to be sensibly cooled to restore an effective dehumidification and sensible cooling potential. A low energy solution is to use an evaporative process. Evaporative cooling of the desiccant can either be a concurrent process alongside dehumidification, taking place on the other side of the dehumidifier exchanger wall, or an evaporative cooling tower can be employed to produce cooled water which transfers coolth to the desiccant solution in a plate heat exchanger. During this process, the desiccant solution is sensibly cooled without moisture addition. Following this the desiccant solution is strong and cool (D1), restoring its dehumidification and sensible cooling potential.

Desiccant based air conditioning systems are primarily powered, for the regeneration process, by thermal energy. If this thermal energy is from solar or waste heat sources, it can significantly reduce both the operating cost and associated CO_2 emissions of the air conditioning process. A recent study by Kozubal et al. (2011) of a desiccant enhanced evaporative air conditioning system demonstrated a 30–90 % reduction in energy demand compared to an equivalent VCS. Liquid desiccant air conditioning is a well suited technology for tri-generation system applications, particularly on a domestic building scale. This is due to their efficient use of low grade thermal energy, minimal electrical requirements and operation at atmospheric pressure.

Next, Sect. 2.3.2 summarises the different contactor designs employed in liquid desiccant air conditioning systems.

2.3.2 Liquid Desiccant Contactor Designs

In liquid desiccant systems, contactors are used to bring the desiccant solution into contact with a process air stream to facilitate dehumidification/cooling or regeneration. The desiccant solution usually flows downward due to gravity across the contacting media. The process air stream either flows in a parallel, counter or most commonly in a cross flow manner across the contactor surface, permitting the transfer of heat and mass; this can either be in a direct or in-direct manner. The dehumidifier and regenerator contactors are of a very similar design as they are essential the same process in reverse.

Table 2.3 summarises the four most commonly used contactor designs used as dehumidifiers/regenerators in liquid desiccant based air conditioning systems. Although the packed bed is the most commonly used design, this thesis employs a semi-permeable micro porous membrane based cross flow contactor. The use of membrane technology brings two significant advantages to the desiccant contacting process: (1) it prevents the inherent disadvantage of desiccant carry

Table 2.3 Summary of liquid desiccant air conditioning contactor designs (Jain and Bansal 2007; Das et al. 2012; Abdel-Salam et al. 2013)

Contactor type	Description	Performance	Advantages	Disadvantages	Diagram
Spray chamber (direct contact)	Desiccant is sprayed into a chamber and air is passed through the chamber		Low cost, simple, low air side pressure drop and compact size	High desiccant carry-over, low latent effectiveness, large desiccant side pressure drop	
Packed bed (direct contact)	Contains packings (random or regular) that are sprinkled with desiccant solution, bringing it into contact with an inlet air stream. Most widely used contactor type	Latent effectiveness = 10–90 %/maximum absolute humidity reduction = 0.02 kg_{vapor}/kg_{dryair}/ air side pressure drops = 10–150 Pa	High latent effectiveness, high rate of heat and mass transfer per unit volume. Honeycomb paper packing can hold desiccant, minimising carry-over	No provision for internal cooling, require high desiccant flow rates, high air side pressure drop	 CELdek honeycomb paper packing core
Wetted wall/falling film (direct contact)	Vertical tube/plate over which desiccant solution flows as a thin film due to gravity. Air is supplied from the sides or below		Low cost, low air side pressure drop, high contact surface area, provision for internal cooling	Difficult to achieve a thin film across the whole plate, desiccant carry-over	
Membrane (in direct contact)	Air/desiccant channels separated by a membrane. Membrane allows transmission of water vapour, but not liquid water, therefore no liquid desiccant carry-over	Latent effectiveness = 20–60 %/maximum absolute humidity reduction = 0.008 kg_{vapor}/kg_{dryair}/ COP_{th} ~0.6–0.8	Zero desiccant carry-over, provision for internal cooling	Additional resistance to heat and mass transfer, difficult to achieve complete wetting over the contactor surface area, improper sealing can lead to desiccant entrainment in supply air	 Counter flow liquid-to-air membrane exchanger

over into the supply air stream (indoor air quality, desiccant losses, duct corrosion), especially when corrosive halide salts are used, and (2) it provides a large contact surface area for the heat and mass transfer between the liquid and air. As a result, membrane based contactors can effectively improve the performance of a liquid desiccant based air conditioning processes. The introduction of a membrane provides additional resistance to heat and mass transfer, therefore much work is focussed on improving this with novel supported liquid membranes (Zhang 2006b). The membrane properties have a significant impact on contactor performance. Zhang (2006b) reports that water vapour transmission increases with increasing pore size and porosity, it decreases with membrane thickness and tortuosity. A good membrane material should be thin ~25 μm, with a pore size of around 0.1 μm, and have a desiccant break through pressure of 140 kPa or greater, and a high water vapour permeability (Kozubal et al. 2011).

The most common working solutions employed in liquid desiccant based air conditioning systems are halide salts, including: lithium chloride (LiCl), calcium chloride ($CaCl_2$) and lithium bromide (LiBr) (Jain and Bansal 2007). The performance of liquid desiccant air conditioning systems employing these solutions is heavily documented in the literature (Jain and Bansal 2007; Lowenstein 2008; Mei and Dai 2008). However, halide salts are extremely corrosive and can cause significant damage to air conditioning equipment. As a result, salts of weak organic acids such as potassium formate ($CHKO_2$) have more recently been adopted (Dwyer 2014; Jradi and Riffat 2014a).

Section 3.2 presents a detailed summary of the LiCl, $CaCl_2$ and $CHKO_2$ solutions, including: dehumidification potential, operating concentrations, physical characterisation and the advantages/disadvantages of their use. Currently, there is limited published and available literature within the public domain regarding the specific physical characterisation of the $CHKO_2$ solution in the required operating range for liquid desiccant air conditioning applications. As a result, Sect. 3.2 presents an assimilation of various sources of data to physically characterise the $CHKO_2$ solution in the required operating range for a liquid desiccant air conditioning application.

Jradi and Riffat (2014a) have numerically investigated a counter flow dehumidifier heat and mass exchanger employing a potassium formate solution with a 0.74 solution mass concentration. The simulations show that the latent and enthalpy effectiveness are directly proportional to the intake air temperature, intake air relative humidity and liquid desiccant flow rate, but inversely proportional to the intake air velocity. A numerical case study is presented demonstrating a 59 % humidity and 62 % enthalpy effectiveness. Qiu and Riffat (2010) have experimentally investigated the feasibility of using potassium formate as a liquid desiccant solution in air conditioning applications. A relative humidity reduction of over 25 % is reported when using a strong solution mass concentration, however it does not effectively dehumidify when the air relative humidity is below 43 %. The authors therefore recommend its use in high humidity applications only.

To facilitate the continuous operation of a liquid desiccant air conditioning system regeneration of the liquid desiccant solution is required to restore its dehumidification potential. Regeneration is most commonly carried out by heating the desiccant solution prior to it entering the contactor; however heating of the regenerator air stream is also possible. Heating creates a higher vapour pressure on the solution side, thus driving a mass transfer to the regenerator airstream. Electric, gas, waste heat and solar energy are all viable options. The use of waste/solar heat sources for solution regeneration tends to increase the energy savings whilst increasing the complexity and cost of the system (Jain and Bansal 2007; Buker and Riffat 2015). To date solar energy has been used in various applications due to solution regeneration temperatures being sub 80 °C and cooling loads being in-phase with available solar radiation (Gommed and Grossman 2007). However, as demonstrated in Sect. 2.4, the hot water outlet temperature from a typical SOFC CHP system is in good thermal agreement with the temperatures required for solution regeneration (Hawkes et al. 2009), demonstrating the potential for the development of a SOFC liquid desiccant tri-generation system. Jradi and Riffat (2014c) state that potassium formate has a lower regeneration temperature requirement (45–60 °C) compared to the halide salts, a statement confirmed in the mathematic modelling in Sect. 3.5.2.

Section 2.3.2 has summarised the contactor designs and introduced the working fluids commonly employed in liquid desiccant air conditioning systems. The advantages of employing an indirect membrane based contactor design along with a potassium formate working solution in the novel tri-generation system has been discussed. Next, Sect. 2.3.3 will review the current literature surrounding liquid desiccant air conditioning systems, with a particular focus on system design and performance.

2.3.3 Liquid Desiccant Air Conditioning Systems and Performance

In this section, a review of liquid desiccant air conditioning systems is provided, with a particular focus on membrane contactors and waste heat driven designs. The application of liquid desiccant in a tri-generation system context is provided in Sect. 2.5.4.

In a liquid desiccant air conditioning system, the desiccant solution needs to be cooled prior to the dehumidification process. This is because it enhances performance through a greater dehumidification potential, removes the latent heat of condensation produced during the dehumidification process, and can provide sensible cooling to the supply air stream. Desiccant solution cooling can be achieved in a variety of ways, including; internal contactor evaporative cooling, use of an internal or external cooling fluid or external pre-cooling. The technique used is dependent on the system and its application. The contactors used in the regeneration process do not require the provision of cooling, but employ an external heating source. Because the cooling achieved in liquid desiccant systems is almost exclusively latent, an additional stage of sensible cooling can be used to lower the supply air temperature (Lowenstein 2008). Liquid desiccant dehumidifiers combined with a VCS or VAS is termed a 'hybrid' system. Another common technological pairing is with evaporative based cooling technology, to create 'stand-alone' systems. Evaporative based cooling compliments liquid desiccant due to its improved performance when operated with dry air and its low electrical requirement (Daou et al. 2006; Ouazia et al. 2009; Kozubal et al. 2011, 2012). Currently, it is reported that hybrid systems are most widely used. This is because of stand-alone system's large geometrical size and system complexity (Jain and Bansal 2007). However, with continual technical improvements to component technology, stand-alone systems are expected to become a major competitor to the hybrid system due to a high COP_{th}, low electrical use and thus environmental impact.

Next, Sects. 2.3.3.1 and 2.3.3.2 presents a review of hybrid and stand-alone liquid desiccant systems respectively. Particular reference to system performance and any operating considerations are made throughout. The summary aims to identify the type of design best suited to the development of a SOFC liquid desiccant tri-generation system.

2.3.3.1 Hybrid Systems

A common and effective hybrid configuration is with a VCS. In this form of hybrid system, the sensible and latent loads can be handled either separately where latent cooling of the air is carried out in the dehumidifier, and air sensible cooling via the VCS cooling coil (Dai et al. 2001). Alternatively, as shown in Fig. 2.6, the loads can be handled simultaneously, where the VCS is used to cool the desiccant solution prior to it entering the contactor where it then cools and dehumidifies the supply air stream (Jain and Bansal 2007). Yadav and Kaushik (1991) found that a 35 % energy saving can be achieved when combining VCS with a liquid desiccant dehumidifier, compared to a straight VCS under normal operating conditions. This is because the VCS does not have to address the latent load and thus the cooling coil can be maintained at a higher ADP, as a result the VCS COP_{el} is higher. In a hybrid system, the VCS condenser's heat output can be used to heat the desiccant solution to facilitate regeneration, generating further energy savings. Compared to a stand-alone (evaporative) system, the hybrid VCS is much more capital intensive and has a higher electrical demand (Mei and Dai 2008).

A hybrid VCS, shown in Fig. 2.7, has been investigated by Dai et al. (2001). The system employs a honeycomb packed bed liquid desiccant dehumidifier, external evaporative coolers and a VCS. The supply air sensible and latent loads are separated. First, the latent load is addressed in the dehumidifier, and then two stages of sensible supply air cooling are carried out, first in the evaporative cooler and then the VCS. A further stage of external evaporative cooling is used to cool the regenerated liquid desiccant solution prior to it re-entering the dehumidifier. The authors reported an improvement in performance and energy savings compared to a conventional straight VCS. The COP_{el} of the system with the VCS

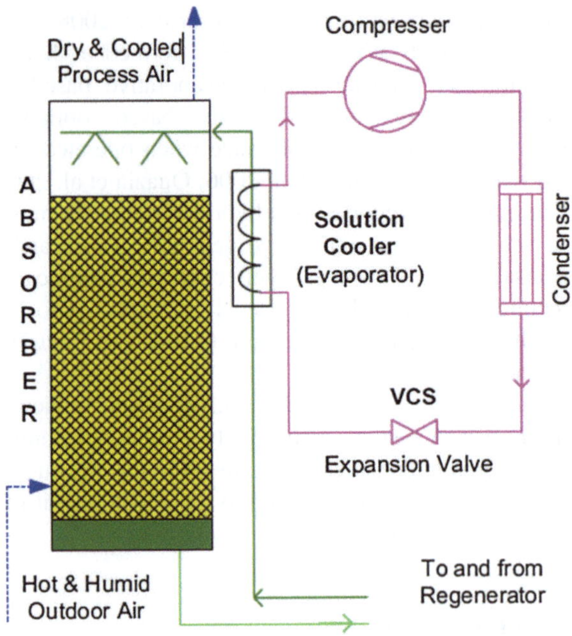

Fig. 2.6 A typical hybrid VCS addressing sensible and latent loads simultaneously (Jain and Bansal 2007)

alone was 2.214; the inclusion of the liquid desiccant contactor increased this to 3.38 and the inclusion of liquid desiccant and evaporative cooler increased this to 3.709; an overall 41 % improvement. The reason for this improvement in performance is that the liquid desiccant and evaporative cooler changed the air state entering the VCS, and thus it only had to address a sensible load. Furthermore, the inclusion of liquid desiccant and evaporative cooling meant that the VCS capacity could be reduced, creating capital cost savings.

Bergero and Chiari (2010) have investigated a membrane based hybrid VCS. The VCS evaporator is used to cool the LiCl desiccant solution prior to it entering the dehumidifier; the supply air is then concurrently cooled and dehumidified in the membrane contactor. The VCS condenser heat output is used to heat the solution to facilitate regeneration. Results from the study show a 60 % energy saving in high latent load applications compared to a straight VCS is possible. Studak and Peterson (1988) investigated LiCl, LiBr, $CaCl_2$ and TEG desiccant solutions for hybrid applications. The solutions were circulated back and forth from the VCS evaporator and condenser. From the study it was found that $CaCl_2$ is the best liquid desiccant; purely based on dehumidification performance.

2.3.3.2 Stand-Alone Systems

Pietruschka et al. (2006) experimentally and theoretically investigated a range of liquid desiccant based air conditioning cycles located on the return air side of a

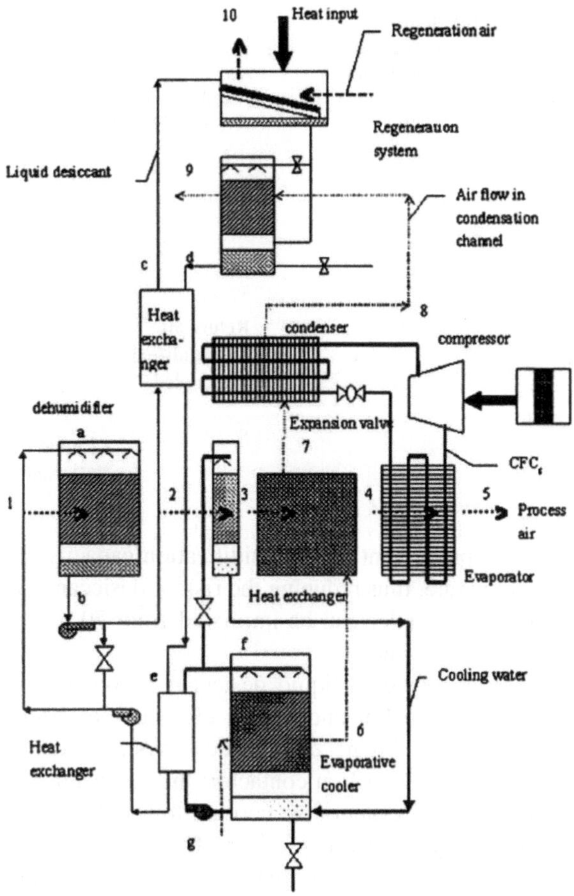

Fig. 2.7 Schematic of the hybrid VCS addressing the sensible and latent loads separately (Dai et al. 2001)

residential building, including a cellulose fibre contactor and an internally evaporatively cooled cross flow contactor. The cellulose contactor achieved better contacting potential, reduced pressure drop and desiccant carry-over compared to the internally cooled design. However, it was found that the best dehumidification performance was obtained with the internally evaporatively cooled contactor. The cellulose fibre contactor cannot be sufficiently internally cooled, and thus its dehumidification potential is limited. Figure 2.8 shows a schematic diagram of the internally cooled contactor. Both LiCl and $CaCl_2$ desiccant solutions were investigated, the LiCl gave 40–50 % higher dehumidification rates. For the summer design condition the system could achieve 886 W of cooling power, generating a supply air temperature of 18.8 °C and absolute humidity of 0.007 kg_{vapour}/kg_{dryair}. The authors reported that if volumetric air flow is tripled from 100 to 300 $m^3\ h^{-1}$ (typical fresh air supply to a domestic dwelling), a modest reduction in absolute humidity of 0.001 kg_{vapour}/kg_{dryair} is achieved. Pietruschka et al. (2006) suggested that if desiccant wetting rates of the plate could be improved from 35 to 60 %, a

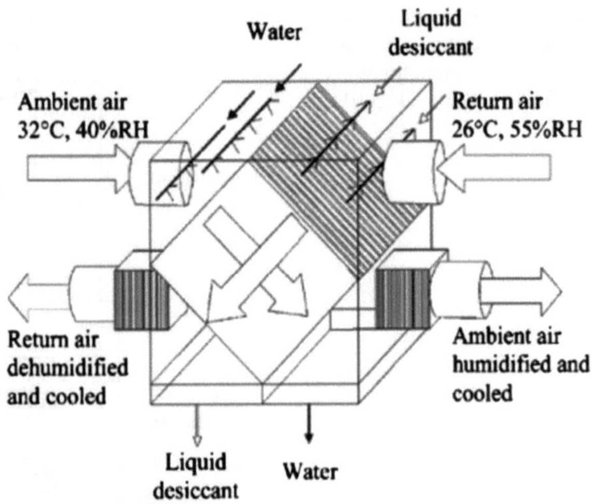

Fig. 2.8 Internally evaporatively cooled cross flow contactor (Pietruschka et al. 2006)

15 % improvement in dehumidification capacity can be achieved at a lower solution flow rate, thus reducing the risk of desiccant carry-over. If wetting rates on the evaporative cooler can be improved from 70 to 76 %, supply air temperatures of 18 °C are possible.

A solar powered liquid desiccant evaporative cooling system for greenhouse applications in hot and humid climates has been investigated by Lychnos and Davies (2012). Figure 2.9 provides a system schematic. The system first dehumidifies air in a falling film contactor; the contactor is internally cooled using cooling water from an evaporative cooling tower. The dehumidified air is then cooled by an additional evaporative cooler. The system can lower the maximum summer air temperature by 15 °C compared to a greenhouse purely conditioned by simple fan ventilation, which is 5–7.5 °C lower than what the conventional evaporative cooler could achieve on its own, indicting the potential of stand-alone liquid desiccant-evaporative systems.

Wang et al. (2009) have experimentally and theoretically investigated a 40 kW liquid desiccant evaporative cooling system for building applications. Figure 2.10 shows a schematic of the system. The system employs a packed bed contactor and uses a LiCl solution. External solution cooling is provided from an evaporative cooling tower. The system is able to treat both sensible and latent loads, with no additional refrigeration required. Inlet ambient air at 35 °C is cooled to the supply requirement of 18 °C. The COP_{th} of the system was found to be 0.8 when the heat source temperate is 70 °C. Because the system can be operated with any heat source above the regeneration temperature it is well suited to solar and waste heat applications i.e. SOFC.

Das et al. (2012) presents an experimental investigation of a solar driven membrane based stand-alone desiccant air conditioning system for hot and humid climates, employing a LiCl solution. The system consists of a dehumidifier, regenerator, solution cooling tower, supply air indirect evaporative cooler (IEC), several

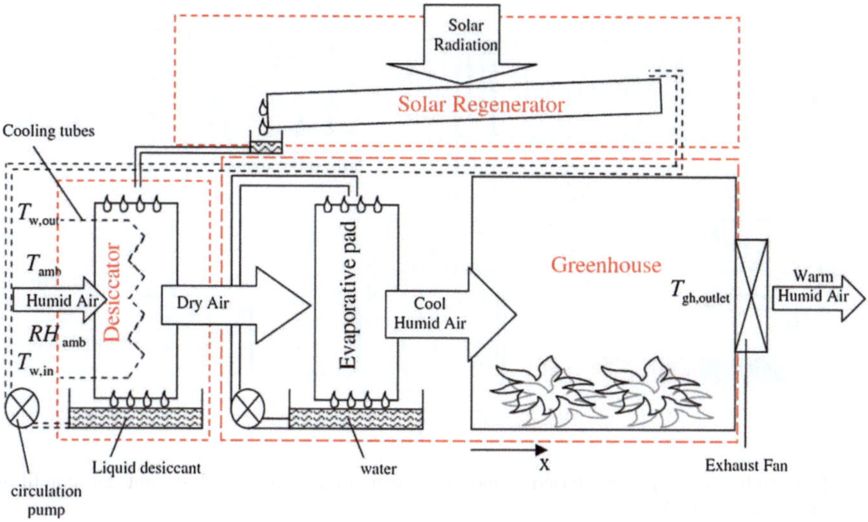

Fig. 2.9 Liquid desiccant-evaporative cooling system for greenhouses (Lychnos and Davies 2012)

Fig. 2.10 Schematic of the desiccant evaporative system (Wang et al. 2009)

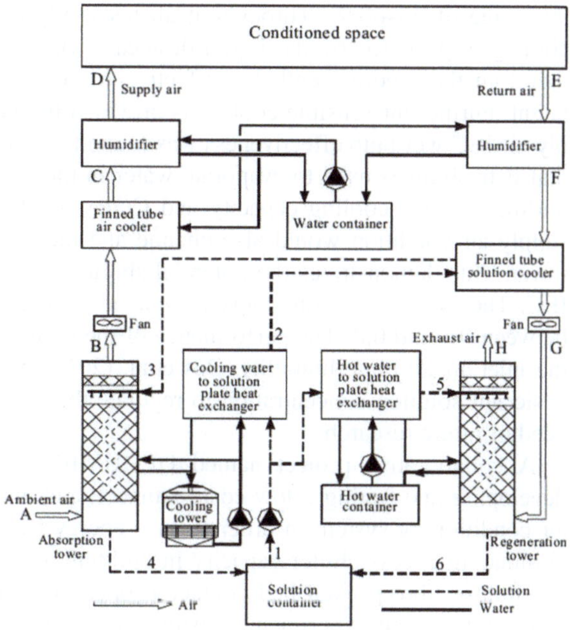

heat exchangers and a solar collector as shown Fig. 2.11. The performance evaluation of the system under a Delhi climate showed that an absolute humidity change of 0.002–0.008 kg_{vapour}/kg_{dryair} and a moisture removal rate of 0.2–1.6 gs^{-1} are

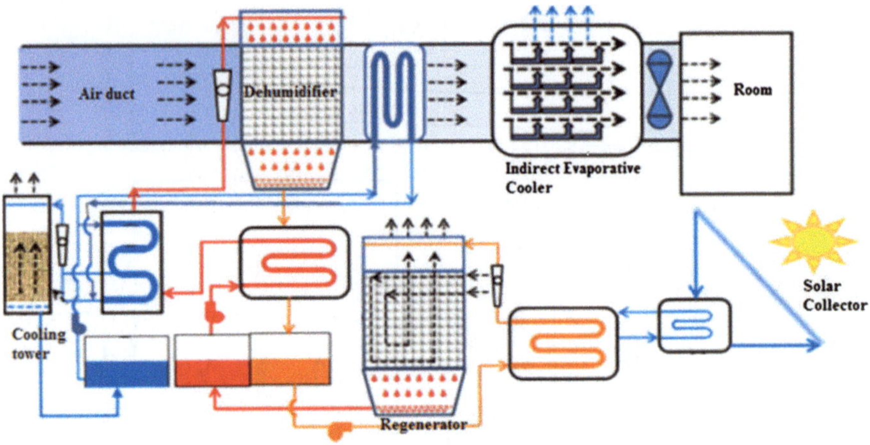

Fig. 2.11 Indirect membrane based stand-alone evaporative cooled desiccant air condition system (Das et al. 2012)

attainable with an inlet air absolute humidity range of 0.01–0.023 kg_{vapour}/kg_{dryair}. The membrane based contactor latent and enthalpy effectiveness is measured in the range of 30–60 %. Throughout all tests only a small amount of air sensible cooling is provided by the liquid desiccant solution in the membrane contactor, therefore the reported enthalpy effectiveness is only marginally higher than the latent. Further air sensible cooling is provided by the second stage IEC. The supply air IEC wet-bulb effectiveness was found to be around 90 %. The authors state that if fresh air is used to evaporate water in the IEC instead of the treated dehumidifier air, the cooling capacity and COP_{th} would almost double. However the supply air condition would also change and the effectiveness of the IEC will be lower. This demonstrates the potential shortcomings of the use of a second stage IEC. The system's cooling capacity ranged from 2.5 to 5.5 kW and the COP_{th} is between 0.4 and 0.8. The performance of the system increases with an increase in the inlet air absolute humidity. Das et al. (2012) believe that the close control of desiccant solution concentration to regulate the supply air condition is an area in need of future research.

A collaborative project named DEHUMID (Bonke 2007) had the goal of developing and testing a low cost, compact, and energy efficient liquid desiccant air conditioning system in order to save energy by reducing compressor size and eliminating excess chiller capacity in building applications. The system employs a cellulose fibre packed bed contactor, operating with a LiCl solution. The data analysis from the experimental testing of the system in a real world application gives encouraging COP_{el} results of around 2.5, with a maximum of 5.9. However, issues encountered included: desiccant leakage and corrosion and insufficient sensor placement creating a lack of data to sufficiently validate mathematical models. A conclusion from the work demonstrates that such systems are only viable in Southern European regions where it is impossible to reduce high indoor air humidity by increasing the supply air flow. Factors requiring further work include material of fabrication/mechanical construction as corrosion and durability issues were encountered, simplification of the control system and the use of hard water in the cooling unit.

Liu (2008) has carried out theoretical and experimental work focussed on a novel heat recovery/liquid desiccant air conditioning system employing a LiCl working solution. The system is composed of an air to air liquid desiccant filmed cellulose fibre heat and mass exchanger combined with a cellulose fibre packed bed contactor driven by solar energy. The complete system is shown in Fig. 2.12.

Conclusions from the experimental work presented by Liu (2008) are summarised with respect to the dehumidifier, regenerator and system performance.
Dehumidifier performance:

- Higher desiccant temperature causes lower heat and mass transfer, and also a higher supply air temperature, thus adequate desiccant pre-cooling is required.
- As desiccant solution supply flow rate is increased, the moisture absorption capacity per litre solution decreases and heat recovery effectiveness increases. Increasing air flow speed results in higher moisture absorption ability, but lower heat recovery effectiveness.
- The use of cellulose fibre packing meant some desiccant carry-over into the supply air stream.

Regenerator performance:

- A higher desiccant temperature results in higher regeneration capacity. A temperature of at least 60 °C was required for a LiCl solution.
- A low desiccant solution flow rate results in higher regeneration capacity and condensed solution concentration.

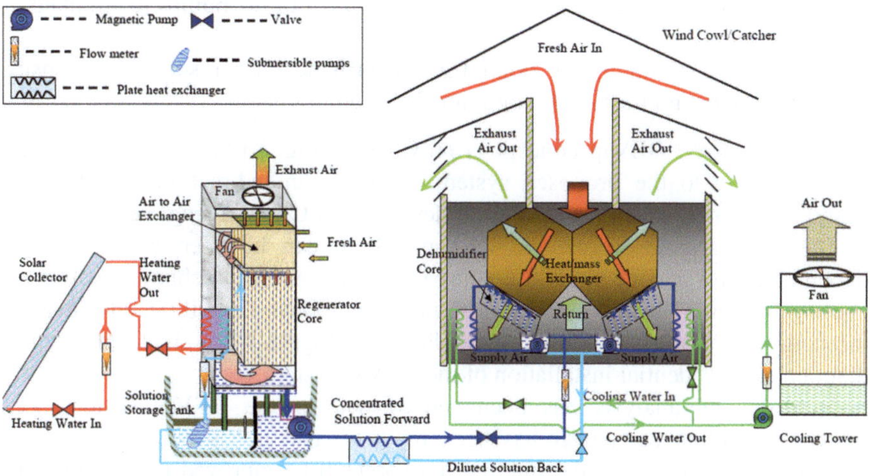

Fig. 2.12 Schematic of the heat recovery/desiccant cooling system (Liu 2008)

- Regeneration capacity and condensed solution mass concentration increases with increasing fresh air flow speed, but then begins to decline, thus there is an optimal air flow speed.
- A dry air stream results in better regeneration capacity.

System performance:

- Cooling capacity and COP_{el} increase with increasing fresh air temperature, therefore this form of system is suitable for hot/humid locations. A maximum COP_{el} of 13 was achieved when utilising solar energy, thus offering the potential for significant energy savings in summer.
- Based on a LiCl solution, when not using renewable/waste heat, the lowest possible concentration of desiccant is preferred as this improves COP_{th}. However when renewable/waste heat is available, the higher the concentration of desiccant results in improved cooling capacity and system COP_{th}.

Lowenstein (2008) and Conde (2007) have identified several issues regarding the use of conventional stand-alone liquid desiccant air conditioning systems, these include; duct corrosion, parasitic energy consumption from fans and pumps, large space requirement and additional requirement of equipment (cooling tower/heat source). Work carried out by Woods and Kozubal (2013) at The National Renewable Energy Laboratory proposes to overcome some of these shortcomings through the design and manufacture of an integrated stand-alone internally cooled dew-point IEC membrane based liquid desiccant contactor (DEVap). The system overcomes the issues listed above by:

- Replacing desiccant spray with a low flow desiccant stream contained behind a semi-permeable membrane, thus reducing pumping power and carry-over.
- Using internal indirect evaporative cooling (IEC) to cool the desiccant and remove the latent heat of condensation, thus reducing the requirement of an external cooling tower and the energy requirement of the pumps to circulate the cooling liquid.
- Using a second stage dew point IEC, to provide further supply air sensible cooling so that no further air conditioning is required.

A diagram of the DEVap concept is provided in Fig. 2.13a. No work has been found relating to the proposed system; combing an indirect membrane based liquid desiccant contactor and internal dew point IEC into a single core. Because of the two separate stages, the DEVap system can be operated as either; a dehumidifier for outside air, IEC cooling only or a combined dehumidification and cooling system. The system demonstrates operational flexibility; advantageous in energy efficient air conditioning applications. Figure 2.13b shows an example diagram of a residential installation of the DEVap system.

Experimental analysis has been carried out for the DEVap system. Zero desiccant solution carry-over has been reported, and it has been demonstrated that that the inclusion of the internal dew point IEC improved dehumidification performance, with an outlet absolute humidity of 0.0084 kg_{vapor}/kg_{dryair} achieved

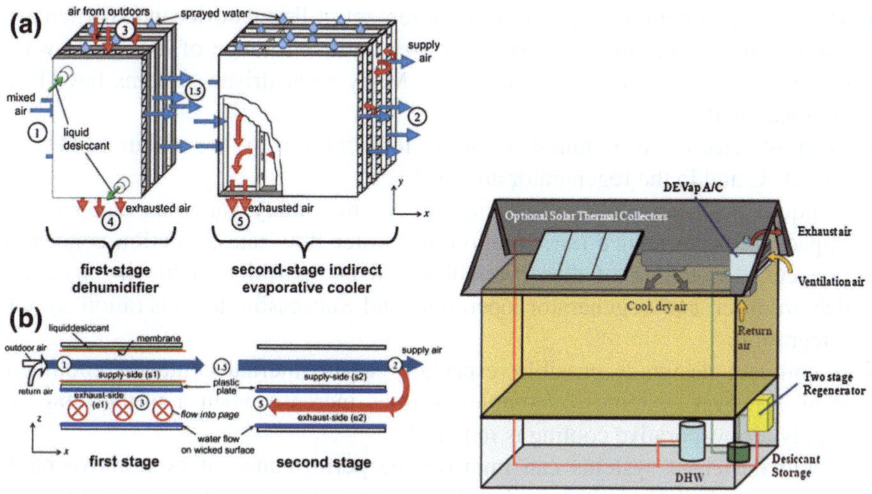

Fig. 2.13 **a** Schematic of the DEVap system, and **b** an example diagram of a domestic installation of the DEVap system (Woods and Kozubal 2013)

compared to 0.0089 kg_{vapor}/kg_{dryair} for the adiabatic test at 40 °C. An effective integrated energy efficiency ratio (EER) of 23.2 was calculated for the combined DEVap system. In comparison, a traditional VCS has an EER of 15.5 (Kozubal et al. 2012). The DEVap concept not only improves the performance of the liquid desiccant dehumidifier, it also acts as an autonomous air conditioner that can adapt to a variety of sensible and latent loads, reduces peak electrical load by 80 % and total energy use by 40–80 % compared to an equivalent VCS (Kozubal et al. 2011).

2.3.4 Liquid Desiccant Air Conditioning Summary

Section 2.3 has provided a review of the literature surrounding liquid desiccant air conditioning. Key conclusions are as follows:

- The use of membrane based contactors eliminates the issues of desiccant carry-over, resulting in improved indoor air quality, reduced plant corrosion and solution losses. Latent (dehumidifier) effectiveness in the 30–60 % range has been reported.
- A small amount of work exists regarding the use of potassium formate as a working fluid in liquid desiccant air conditioning applications, demonstrating good dehumidification capacity, particularly in high humidity applications. There is limited published and available literature within the public domain regarding the specific physical characterisation of the $CHKO_2$ solution in the required operating range for liquid desiccant air conditioning applications.

- The low regeneration temperatures requirement of liquid desiccants, particularly potassium formate, means it can effectively utilise a range of low grade waste or renewable heat sources i.e. fuel cell. Many solar driven systems have been demonstrated.
- In most cases the required solution temperature in the dehumidifier are 15–20 °C and in the regenerator are 45–55 °C.
- A liquid desiccant air conditioning system has many variables that have an impact on performance (air, solution and water flow rates, solution concentration etc.). Matching of these variables will be critical to achieving balanced dehumidifier and regenerator operation and successful tri-generation system integration.
- A compact design is possible, especially when internal contactor cooling is used, however system size/complexity does increase when a further stage of supply air evaporative cooling is required.
- Liquid desiccant systems can improve the performance of evaporative based coolers, particularly in hot and humid climates. However when using IEC, consideration should be given to the process air requirement.
- Hybrid desiccant systems demonstrate good overall COP_{el}, improved VCS performance and a more compact form; however the increased electrical consumption compared to the stand-alone systems is undesirable for a tri-generation system application.
- Low electrical requirement and high COP_{th} makes stand-alone evaporative based systems an attractive option for effective waste heat utilisation. Operating COP_{th} of up to 0.8 has been reported, demonstrating the potential for tri-generation system integration.
- The effectiveness of stand-alone liquid desiccant evaporative systems is dependent upon inlet air conditions; therefore the geographical location needs careful consideration, and for the tri-generation system control of the inlet environmental conditions may be required.

It is apparent that a membrane based stand-alone liquid desiccant air conditioning system is the most appropriate system in the development of an efficient and effective tri-generation system.

Next, Sect. 2.4 discusses the use of fuel cell CHP in (domestic) built environment applications.

2.4 Combined Heat and Power

This section discusses the use of fuel cell CHP in (domestic) built environment applications. Tri-generation can be considered a sub-set of CHP because the prime mover technology is the same and the operating aims alike. Thus, the discussions presented regarding fuel cell CHP in built environment applications provides a good basis to discuss tri-generation systems in Sect. 2.5.

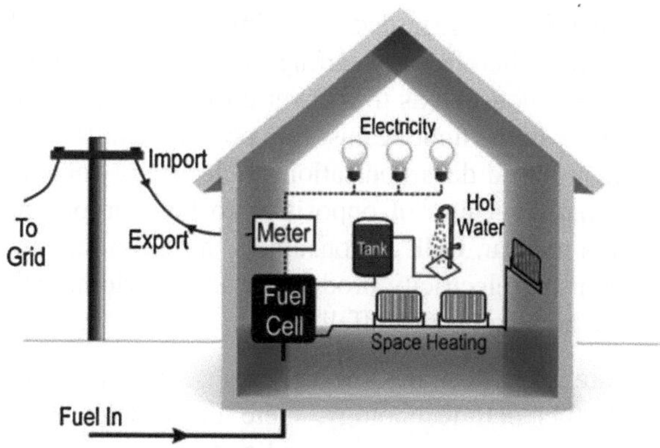

Fig. 2.14 CHP in a domestic building (Hawkes et al. 2009)

CHP is defined as the generation of heat and power from a single fuel source, with the view to using both products. Figure 2.14 shows a typical domestic CHP system configuration. Fuel is supplied to the prime mover technology from the central network to produce electrical power and in the process generates heat. The electricity is used directly in the home, and if grid interactive, can be imported or exported as required. The heat produced in the electrical generation process is recovered and used in applications such as space heating or domestic hot water. By consuming this heat, system efficiency can be elevated from as low as 20 % to over 90 % depending on the prime mover technology and the extent of waste heat utilisation (Choudhury et al. 2013).

In built environment applications elevated system efficiency results in reduced primary energy demand, leading to decreased emissions and running cost for the consumer. However, as Beaussoleil-Morrison (2008) states, if the thermal output of the CHP system cannot be fully utilised, then the system cannot expect to deliver a net benefit relative to grid electricity and a highly efficient condensing boiler. Therefore accurate building energy load assessments and sizing of the CHP unit is essential (El-Gohary 2013). Currently, The Carbon Trust defines domestic scale CHP as any system up to 3 kWe. Because CHP and tri-generation systems produce electricity at point of use, they are often referred to as decentralised energy conversion devices. Decentralised energy conversion in the built environment has many advantages associated with it compared to traditional centralised conversion, including:

- Improved system efficiency, otherwise wasted heat is utilised (heating or cooling), therefore system efficiency can be elevated from as low as 30–50 % in central power stations to around 70–90 % (Choudhury et al. 2013).
- Decentralised energy conversion with CHP significantly reduces transmission losses, which account for 6–24 % in the European transmission network (Peht et al. 2006).
- Improved system efficiency and greater fuel utilisation leads to reduced primary energy demands, resulting in cuts to CO_2 emissions and operating costs (Fubara et al. 2014).

- Electricity is regarded as having an economic value of roughly three times that of gas. Therefore converting lower cost gas (common fuel in CHP) to electricity allows households to recover cost and reduce energy bills. This is an important factor in the fight against fuel poverty (Staffell 2009).
- Centralised decarbonisation of electricity generation in many countries is problematic because of opposition to low carbon technologies such as renewables and nuclear. CHP in consumers' homes offers an option to assist in both decarbonising electricity production and providing energy saving benefits directly to the home owner (Staffell 2009).

Figure 2.15 illustrates the benefits of generating heat and power at point of use with fuel cell technology. Note the efficiencies are illustrative and will vary depending on the systems used and countries involved. In a tri-generation system the CHP heat output is simply used to provide cooling in summer.

Currently, there are three key technologies used as prime movers in CHP and tri-generation systems in the built environment. All are at varying levels of commercial and technological maturity; internal combustion engine (ICE), Stirling engine (SE)—both combustion based technologies, and fuel cells. The performance and operational characteristics of these three technologies are summarised in Table 2.4. The data presented in Table 2.4 is indicative and used for comparative purposes. The data is quoted for CHP systems in the range up to 5 kWe. It should be noted that CHP performance, primarily electrical efficiency, is highly dependent upon fuel used, engine type, engine size and running mode. Larger capacity ICE and SE based systems will have higher electrical efficiencies (Bianchi et al.

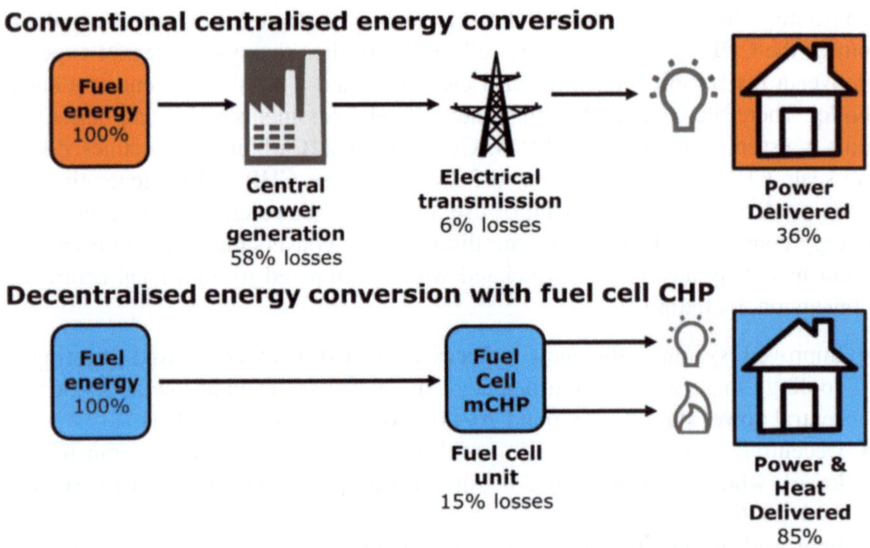

Fig. 2.15 Scheme of the advantages of decentralised energy conversion with fuel cell CHP compared to conventional centralised energy conversion

Table 2.4 Built environment CHP technologies (Wu and Wang 2006; Hawkes et al. 2009; Staffell 2009; Vourliotakis et al. 2010; Steinberger-Wilckens 2013)

	Internal combustion engine	Stirling engine	Fuel cell
Capacity (kW_e)	1–5	1–5	0.7–5
Electrical efficiency (%)	20–30	10–20	PEMFC—30 to 40 SOFC 40 to 60
Overall system efficiency (%)	Up to 90	Up to 95	Up to 85
Heat to power ratio	3	8–10	PEMFC—2 SOFC—0.5 to 1
Able to vary output	No	No	PEMFC—Yes SOFC—No
Fuel used	Gas, biogas, liquid fuels	Gas, biogas, butane	Hydrocarbon, hydrogen
Noise	Loud	Fair	Quiet
Maturity	High	Fair	Low
Companies	Vaillant ecoPower	EHE Wispergen	Baxi, CFCL

2012). It is evident that fuel cell technology has some clear operational advantages, particularly when operating at the less than 5 kWe scale in the domestic built environment. Advantages include; higher electrical efficiencies, low heat to power ratio and near silent operation. However, the relative infancy of fuel cell technology has limited their extensive application and market involvement to date.

Because of the low electrical efficiency and correspondingly high thermal output of the internal combustion and Stirling engine, they should only operate when their thermal output can be fully utilised, otherwise the CHP system cannot expect to deliver a net benefit relative to grid electricity and a highly efficient condensing boiler (Beaussoleil-Morrison 2008). Fuel cells, however, with their higher electrical efficiency have much lower heat to power ratios; therefore their operation can be largely independent of thermal demand, making them a well suited technology for domestic CHP applications (Hawkes et al. 2009). Fuel cells can be operated in an electrically led manner, thus providing increased net benefit to the user. SOFCs have lower heat to power ratios compared to PEMFC technology, making them a well suited technology for domestic applications.

Figure 2.16 shows the annual CO_2 savings per kW_e capacity of various CHP systems compared to a grid electricity and boiler alternative (Maghanki et al. 2013). The data shows the SOFC system has the best emission performance, with the PEMFC system second, excluding the 5 kW_e gas engine. The high electrical efficiency seen in the SOFC CHP system means it can generate large emission savings due to a greater displacement of carbon intensive centralised electrical generation. Although the PEMFC system has a high electrical efficiencies in the 1–5 kW_e range, as the electrical capacity of the gas engine increases to 5 kW_e, its efficiency will also increase, therefore improving the emission performance, and in this case surpassing the PEMFC performance.

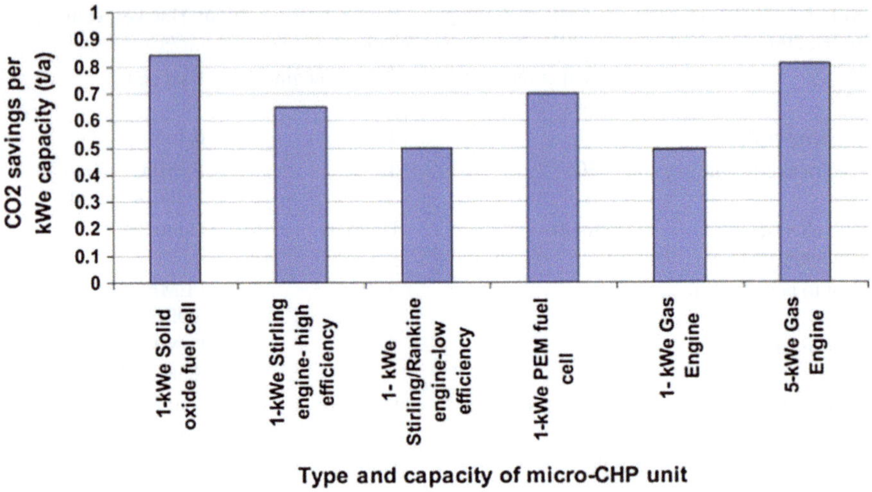

Fig. 2.16 Annual micro CHP CO_2 savings compared to grid electricity and boiler alternatives (Kuhn et al. 2008; Maghanki et al. 2013)

Figure 2.17 shows that according to a report by Delta-ee energy consultants, fuel cell CHP systems for domestic applications outsold conventional combustion based systems for the first time in 2012, accounting for 64 % of global sales, illustrating a major shift in the domestic CHP market on account of fuel cells clear operational advantages (FCT 2013). Growth is being driven by Japan and to a lesser extent Germany, which together account for more than 90 % of yearly sales (Bradley 2013). However as Steinberger-Wilckens (2013) states, for fuel cells to produce a marked effect on the stationary market they need to match and surpass the performance of current CHP technologies such as SEs and ICEs.

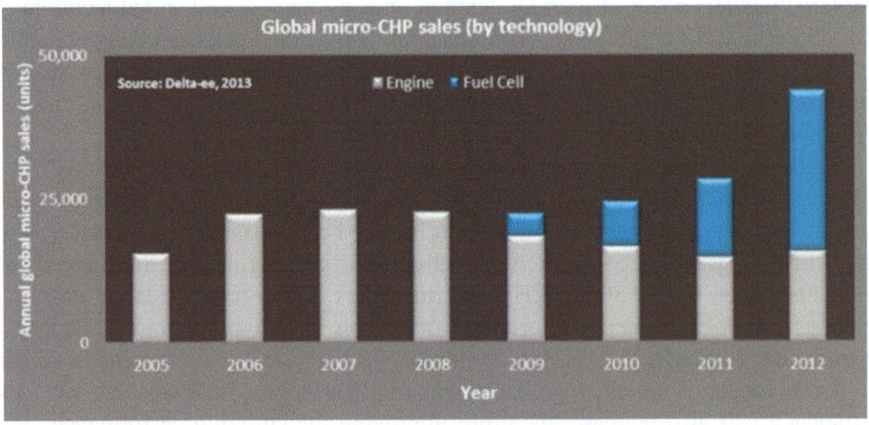

Fig. 2.17 Global micro-CHP sales by technology (Bradley 2013)

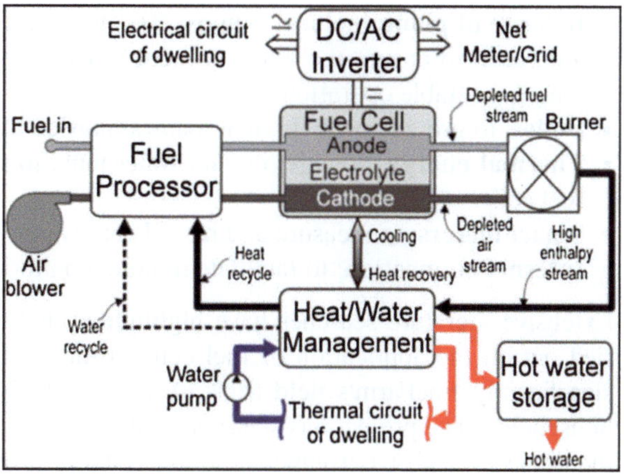

Fig. 2.18 A typical fuel cell CHP system for domestic building applications (Hawkes et al. 2009)

Steinberger-Wilckens proposes that meaningful performance indicators include; amount of CO_2/fossil fuel avoided through the use of fuel cell technology or the total and electrical efficiency of the fuel cell system.

A schematic diagram of a typical fuel cell CHP system for domestic building applications is shown in Fig. 2.18. A fuel cell in principle is very simple, requiring few parts (even less moving), resulting in near silent operation and little maintenance required. However, in order to operate a fuel cell system i.e. a load supplied by a fuel cell, many auxiliary devices and interconnections are needed for both the correct operation of the fuel cell and the delivery of heat and power to the load. Some of these auxiliary devices have a power demand, thus they are a parasitic load on the system. The electrical efficiency of the entire system can be between a fifth and a third less than the quoted stack efficiency due to these parasitic loads (Hawkes et al. 2009). Auxiliary equipment also contributes to increased noise, vibrations and maintenance. A 1–5 kW_e fuel cell CHP system can expect to produce 0–55 dB, whereas an ICE is around 95 dB (Hawkes et al. 2009).

The following list provides an explanation of the components found in the fuel cell CHP system shown in Fig. 2.18. In a tri-generation application, the thermal output is simply used in a heat driven cooling cycle.

- **Fuel cell stack**: where hydrogen and oxygen are combined to produce electricity, heat and water.
- **Fuel processor**: converts a hydrocarbon fuel such as natural gas into hydrogen and CO_2 (not required in an SOFC system).
- **Inverter, grid tie and power electronics**: converts the DC electrical output of the fuel cell stack into AC electrical power to either serve the buildings energy demands or to be fed back to the grid.
- **Waste heat recovery (WHR) system**: used to recover heat from the fuel cell in order to improve; (a) the performance of the fuel cell stack and (b) the environmental performance i.e. make it a CHP system. In a domestic application flow temperatures of around 50–70 °C are common (Ellamla et al. 2015).

- **Balance of plant**: includes pumps, fans, valves, sensors, piping and control system, used to ensure the whole system functions in a safe, efficient manner for long term stable operation.
- **Boiler**: to provide peak thermal loads alongside the fuel cell.
- **Thermal energy storage**: i.e. hot water tank, to store the thermal output of the fuel cell.
- **Smart meters**: to measure and record energy conversion and consumption.
- **Internet connection**: to facilitate remote monitoring and data acquisition.

Extensive literature searches have highlighted that successful work has been carried out on the application of fuel cells in the domestic built environment; most significantly EneFarm's field trials of a 1 kW_e PEMFC CHP system in Japanese households. The project has spear-headed the wider, global use of fuel cell CHP in the domestic built environment and the systems have confirmed annual CO_2 emission reductions in the order of 750–1250 kg per annum are possible, illustrating the significant potential fuel cells have in assisting decarbonisation of the future domestic built environment (Elmer et al. 2015b). Ren and Gao (2010) have investigated a gas engine and fuel cell for domestic CHP applications in Japan. Results indicated that the fuel cell offered superior economic and environmental performance in such applications.

Section 2.4 has discussed the use of fuel cell CHP in (domestic) built environment applications. Next, Sect. 2.5 provides a review of the literature surrounding tri-generation systems, particularly those employing fuel cell and liquid desiccant technology.

2.5 Tri-generation Systems

The majority of tri-generation systems for building applications use the thermal energy rejected by the electrical generator to produce a useful cooling output. An extensive review of the literature has shown that both VAS and desiccant air conditioning systems have been used extensively (Badami and Portoraro 2009). The most common technological pairing has been found to be an ICE with a VAS (Minciuc et al. 2003; Al-Sulaiman et al. 2011; Wu et al. 2014). No research publications have been found describing a SOFC or even fuel cell based liquid desiccant tri-generation system. Due to the lack of work regarding the proposed system, this section of the literature review has been split into four key sections:

(1) Domestic scale fuel cell tri-generation systems in Sect. 2.5.1
(2) Large scale fuel cell tri-generation systems in Sect. 2.5.2
(3) Domestic scale combustion based tri-generation systems in Sect. 2.5.3
(4) Liquid desiccant based tri-generation systems in Sect. 2.5.4

This approach has been adopted to achieve a broader context of research, with the aim of drawing conclusions that are relevant to the specific SOFC liquid desiccant tri-generation system. Discussions will cover; what has been done to date, transferable lessons from equivalent systems or applications, system design and performance, operating issues and evaluation techniques. The advantages fuel cell technology can provide over conventional combustion based technology in tri-generation system applications is highlighted and the future of fuel cell tri-generation systems in the built environment discussed.

As previously stated, the proposed novel tri-generation system is applicable to a range of building scales. However, the thesis is focussing on the development of a system of 1.5 kW_e, and therefore suitable for domestic building use. As a result, the literature review will, where possible, focus on this building type. The attributes that make fuel cells an attractive option for tri-generation system applications have been discussed in Sects. 2.2 and 2.4. Fuel cell CHP systems, of all scales, operating in building applications are well documented in the literature (Hawkes et al. 2009; Ellamla et al. 2015). A fuel cell tri-generation system is in principle very similar to a fuel cell CHP system, the thermal output is simply used in a heat driven cooling cycle instead of serving building heating demands.

2.5.1 Domestic Scale Fuel Cell Tri-generation Systems

This section will examine fuel cell based tri-generation systems for domestic building applications. Tri-generation is a well-known technology for energy conservation in commercial and industrial applications. However, very limited work has been completed for tri-generation systems in domestic building applications (Míguez et al. 2004; Huangfu et al. 2007). Kong et al. (2005) states that the concept of tri-generation systems for individual domestic buildings has only be thought reasonable with the more recent development of heat driven cooling technologies with capacities of <10 kW that can operate on low-grade thermal energy (60–90 °C). Huangfu et al. (2007) believes the main obstacles to any type of domestic scale tri-generation systems is the high initial cost and complexity of optimum matching of different parts of the system i.e. prime mover and heat driven cooling. Other commonly referenced obstacles include; system size and complexity. However with recent advances in liquid desiccant based air conditioners for small scale residential applications; the development of a fuel cell tri-generation system in homes is possible (Pilatowsky et al. 2011).

Míguez et al. (2004) and Porteiro (2004) state that the introduction of tri-generation systems to the domestic built environment requires the core of the system, the CHP unit, to be compact, cost efficient and easily installed. With the market introduction of commercially available fuel cell CHP units such as BlueGEN and GAMMA, the possibility of fuel cell tri-generation systems for domestic homes is becoming more realistic. Gigliucci et al. (2004) have conducted extensive work on fuel cell CHP systems in domestic built environment applications, in particular their thermal management. Conclusions drawn from their work include; the abilities of delivering the waste heat to a useful heat sink (tri-generation applications will increase this), capacity to vary the heat to power ratio and electrical output during operation are all critical for the full potential of the fuel cell device to be realised. Fuel cells with their lower heat to power ratios show great promise in terms of total thermal energy utilisation, illustrating why fuel cell technology has been highlighted as an obvious candidate for tri-generation domestic built environment applications (Deng et al. 2011; Pilatowsky et al. 2011).

Bhatti et al. (2010) present a patent publication for a SOFC assisted air conditioning system based on solid desiccant and evaporative cooling technology. The aim of the system is to provide both comfort cooling and heating. The authors state that the most common technology to provide heating and cooling is a reversible heat pump; however these use electric power to drive the compressor. Desiccant enhanced systems are a low energy alternative to this. The thermal energy from the SOFC is used to regenerate a solid desiccant wheel which is used to address the latent load, whilst an evaporative cooler is used to address the sensible load. No performance data is available, however it is expected the system will provide a high system efficiency and energy reduction in comparison to a separate vapour compression, boiler and grid electricity system.

Pilatowsky et al. (2007) have carried out simulation based work on a 1 kW_e PEMFC coupled to a VAS. The simulations were completed to determine the optimum operating conditions of the VAS during the co-generation process; primarily the cooling capacity at maximum power output from the PEMFC. The VAS was operated with a mono-methylamine water solution, with low vapour generation temperatures of around 80 °C, ideal for PEMFC CHP applications. Results show that the co-generation process increases total efficiency of the PEMFC system, illustrating the feasibility of using fuel cells in small scale tri-generation system applications. Results from the simulation include; (1) The COP_{th} of the absorption cooler increases as the generation temperature (from the fuel cell) increases, reaching a maximum value and then dropping off (2) The cooling power increases with an increase in the electrical power from the PEMFC and evaporation temperature of the absorption cooler, but decreases with an increase in generation temperature, and (3) The co-generation efficiency is almost independent of the electrical power, but strongly dependant on evaporation and generation temperatures of the absorption cooler.

A 1.5 kW_e low temperature SOFC liquid desiccant tri-generation system has been proposed by Riffat (2012) as an environmentally friendly way of providing heating, cooling and electricity to low carbon domestic buildings. It has been predicted that the described system running on hydrogen from the natural gas network will result in a 70 % CO_2 emission reduction compared to a traditional energy conversion system comprising separate condensate power plant, boiler and compressor driven cooling unit. The tri-generation system will be constructed at The University of Nottingham as part of an EU funded TriSOFC project, Agreement no. 303454 (Riffat 2012).

Section 2.5.1 has focussed on domestic building scale fuel cell tri-generation systems. Domestic scale fuel cell tri-generation shows great potential with respect to energy utilisation, security and operational costs. However, currently, an extensive gap in the literature exists regarding such systems. Sonar et al. (2014) states that developmental barriers are more pronounced in smaller (micro) scale systems than they are in larger applications. This is primarily due to the fuel cell capital cost. However, the technical complexity of the system, developing heat driven cooling technologies of less than 10 kW and a lack of supporting services and incentives for tri-generation systems in the domestic built environment are also key barriers. As a result, the work that has been completed to date has mainly focussed on simulation studies. Ellamla et al. (2015) believe that doubling production volumes of fuel cell CHP units in domestic built environment applications will bring about a 25 % reduction in cost, therefore increasing the feasibility of adopting fuel cell tri-generation in the domestic built environment.

Next, Sect. 2.5.2 summarises the literature surrounding large scale fuel cell tri-generation systems.

2.5.2 Large Scale Fuel Cell Tri-generation Systems

Yu et al. (2011) have numerically investigated a tri-generation system incorporating a SOFC and a double-effect water/lithium bromide VAS, high total system efficiencies of 84 % or more were reported by the authors, illustrating the benefits of tri-generation systems in applications where heating, cooling and power are required. Margalef and Samuelsen (2010) numerically examined a 300 kW MCFC VAS tri-generation system, achieving an overall system efficiency of 72 %. The pairing of two off the shelf technologies for tri-generation system construction was shown to be problematic. Margalef and Samuelsen (2010) state that the MCFC and VAS chosen for the tri-generation system were close, but not an ideal match. The fuel cell exhaust gas temperature was higher than the inlet temperature specified for the VAS and the exhaust flow rate was not sufficient to achieve the potential heat recovery within the VAS heat exchanger. Therefore, two strategies to overcome this were presented (1) blending the fuel cell exhaust gas with ambient air, and (2) mixing the fuel cell exhaust gases with a fraction of the chiller exhaust gas. Both options worked, however the second option yielded better performance and was thus adopted. Al-Sulaiman et al. (2010) presents an energy analysis of a tri-generation plant incorporating a 520 kW SOFC, organic Rankine cycle, heat exchanger and single effect VAS. The investigation showed that by incorporating the cooling cycle system efficiency is improved by 22 % compared to just having the SOFC and organic Rankine cycle running together. A maximum tri-generation efficiency of 74 % has been achieved.

Fong and Lee (2014) have investigated a SOFC tri-generation system for high-rise buildings in a hot and humid climate. The study focussed on two sizing options. (1) Full SOFC, where the system was sized to peak loads, therefore requiring no grid import, and (2) partial SOFC, where the system was sized such that peak loads were met by the SOFC and grid, however over the course of one year the system maintains a net zero grid import. The environmental and energetic performance of both cases was investigated over the course of a year using a dynamic simulation. The full and the partial SOFC systems generate a 51.4 and 23.9 % carbon emission saving respectively, and a 7.1 and 2.8 % electricity saving respectively. The full SOFC tri-generation system showed the best environmental and energetic performance due to the partial SOFC systems requirement of grid electricity. However the economics of sizing the tri-generation system to meet peak load capacity was not investigated. This should be the focus of future work; to determine the optimum sizing and operating strategies for the studied systems.

Zink et al. (2007) have examined a 110 kW SOFC based tri-generation system employing a VAS. Results show that total system efficiency can reach 87 % or more and that the combined system shows great advantages both technically and environmentally over other current CHP and tri-generation systems. Since many heat driven cooling technologies such as VAS are already commercially available, the viability of fuel cell tri-generation systems heavily depends on the commercialisation of the fuel cell CHP system. With recent trends indicating a reduction in the cost of fuel cell based CHP systems (FCT 2013), the authors believe fuel cell tri-generation systems will be commercially viable within the decade. Darwish (2007) have investigated a tri-generation PAFC system to meet high air conditioning loads in a large building in Kuwait. The PAFCs thermal (105 kW) and electrical outputs (200 kW) are used in a VAS and VCS respectively. Cold water storage has been used to allow the fuel cells limited output to meet both average and peak air conditioning loads. It was proposed that the economy of the system only becomes feasible once the fuel cells capital cost drops below 2000 $ kWe^{-1} (~1300 £ kW_e^{-1}). Fuel cell based tri-generation systems show great promise in terms of energy savings. However, the economics of such systems are not yet feasible due to the current price of the technology, a finding Malico et al. (2009) support in their study of a SOFC tri-generation system in a hospital building.

Section 2.5.2 has focussed on large commercial/industrial scale fuel cell tri-generation applications. The literature presented demonstrates good performance for fuel cell tri-generation systems in terms of tri-generation system efficiency, primary energy demand reduction, and associated CO_2 emissions/operational costs. However, issues regarding the accurate pairing of prime mover and cooling technologies needs careful consideration to ensure effective system operation. In large scale applications, the economic viability of any tri-generation system, but particularly fuel cell, lies with the capital cost of the fuel cell, not the cooling technology, which is already at a level commensurate with economic viability. Darwish (2007) estimates a fuel cell cost of 2000 $ kWe^{-1} or below is required to permit the wider use of fuel cell tri-generation systems, a cost target figure Staffell and Green (2013) believe is possible by 2020. Finally, recent advances in low temperature SOFC technology using ceria-carbonate two or multi-phase nanocomposite has illustrated that high electrochemical performance (1.2 W cm^{-2}) can still be achieved at reduced temperatures (500 °C). These developments mean SOFC stacks could be manufactured for prices below 450 $ kWe^{-1} (~300 £ kW_e^{-1}) (Fan 2013), demonstrating the potential for future fuel cell tri-generation systems across all building scales and applications.

Next, Sect. 2.5.3 presents a review of the literature surround domestic scale combustion based tri-generation systems.

2.5.3 Domestic Scale Combustion Based Tri-generation Systems

This section will review combustion based tri-generation systems for domestic built environment applications. Due to the limited breadth of literature surround fuel cell based systems demonstrated in Sect. 2.5.1, this approach will both provide conclusions that may be applicable to a fuel cell based system and also highlight the operational benefits of adopting fuel cell technology in tri-generation systems at a domestic building scale.

Khatri et al. (2010) state that the use of tri-generation systems of less than 10 kWe for domestic applications is feasible and an effective technique to utilise resources efficiently. The authors tested a 3.7 kW compression-ignition engine tri-generation system in a laboratory. Results showed the thermal efficiency in tri-generation mode reached 86.2 %, compared to 33.7 % in single generation mode. This illustrates great promise for domestic scale tri-generation systems. Optimisation and energy utilisation is not a large issue in theoretical or laboratory based projects, thus high system efficiencies have been reported in various sources of literature. In a real world working environment, effective energy utilisation could pose a serious challenge to the system, specifically how to maximise the utilisation of the energy outputs of the system in order to improve system efficiency. If this cannot be addressed it will have a large impact on the final feasibility of the system. Future work, such as that presented below, should focus on this consideration.

An experimental, tri-generation system for domestic buildings, employing a combustion based prime mover is summarised below. This review is used is to illustrate the benefits of, and any operational considerations for, the tri-generation concept in domestic buildings, and to demonstrate the added value fuel cell technology can bring to the system. Kong et al. (2005) have developed a novel tri-generation system aimed at domestic built environment applications. The system comprises a 12 kW electrical/28 kW thermal ICE and a 9 kW VAS. Initial experimental studies carried out by Kong et al. (2005) show the system can achieve an overall thermal and electrical efficiency of over 70 % under real life operating conditions. Huangfu et al. (2007) have carried out further investigations into the system developed by Kong et al. (2005). The analysis has focussed on the systems economic, energetic and exergetic performance. The payback period and net present value (NPV) of the system show the greatest sensitivity to changes in the price of natural gas. Due to the fluctuating nature of natural gas prices in many counties today, this sensitivity highlights the delicate nature of the economic feasibility of tri-generation systems operating on natural gas in domestic applications. Furthermore, it is an obvious and extremely important consideration for the economic feasibility of fuel cell tri-generation systems, on account of the fact that the majority of current fuel cell CHP systems require natural gas to operate (BlueGEN, Baxi, EneFarm).

The system described by Kong et al. (2005) was energetically evaluated in terms of its primary energy ratio (PER); the ratio of required energy output of the system to the primary energy demand. The higher the PER, the superior the system is with regards to energy consumption (Havelský 1999). The tri-generation system outperformed the comparison independent system of grid electricity, boiler and electrical cooling, when operating as a CHP device, however when operating as a combined cooling and power system (CCP); the independent system had a higher PER. This is because in CHP mode the waste heat is used effectively. In CCP mode, thermal energy is not utilised so effectively due to the low COP_{th} of the VAS, which was around ~0.3. Thus, the greater the heat demand that can be met by the tri-generation system, the higher the PER achieved, and therefore the greater the energy savings the tri-generation system can expect to achieve in comparison to the independent system. For future tri-generation systems, operating with or without fuel cell technology, the energetic performance of the heat driven cooling technology needs to be higher in order for the system to be competitive in all types of operation i.e. CHP and CCP. Open cycle absorption systems, also known as desiccant air conditioners, have been successfully used in tri-generation applications where cooling capacities of less than 10 kW are required. Performance figures for these systems show the potential for a COP_{th} in the range of 0.7–1.5 (Liu et al. 2004), an 80 % improvement on the adsorption system used in the system described by Kong et al. (2005). Higher COP_{th} would make the tri-generation system described competitive with the independent system in CCP mode. Desiccant air conditioning systems could therefore be an appropriate and effective technological option for both combustion and fuel cell based tri-generation systems operating in domestic building applications.

Finally, an influence factor analysis was conducted to see the impact electrical efficiency, heat recovery effectiveness, cooling COP and the ratio of recovered heat for driving the VAS had on the PER. It was found that the electrical efficiency of the prime mover device had the largest impact on the PER. In addition to this, an exergetic analysis was conducted, with the authors concluding that the complete tri-generation system would benefit from operation that maximises electrical efficiency. Therefore the authors concluded that the tri-generation system should be operated at conditions that maximise its electrical efficiency in order to bring about the best energetic and exergetic performance. Fuel cells, particularly SOFCs have much higher electrical efficiencies compared to both the ICE and SE. Fuel cells are therefore a key technological option for maximising the energetic, exergetic and economic performance of tri-generation systems in domestic built environment applications.

Section 2.5.3 has discussed the use of combustion based tri-generation systems in domestic built environment applications, and how the use of fuel cell and liquid desiccant technology can potentially enhance the performance of such systems. The reviewed work shows high tri-generation system efficiency is attainable;

however, high COP_{th}, maximised electrical efficiency and low/stable gas prices are all required in order to make tri-generation a viable option for domestic built environment applications. Fuel cells, particularly SOFC, with their high operating electrical efficiency demonstrate great potential for the replacement of combustion based technologies in tri-generation system applications of this scale. Similarly, liquid desiccant systems at a domestic building scale have higher COP_{th} values compared to VAS and thus have the potential to improve the performance in tri-generation cooling mode.

Next, Sect. 2.5.4 summarises the literature surrounding liquid desiccant based tri-generation systems.

2.5.4 Liquid Desiccant Based Tri-generation Systems

This section will examine tri-generation systems employing liquid desiccant air conditioning technology.

A system employing an ICE and VAS is provided below; the work illustrates the benefits a liquid desiccant air conditioning system can bring to a tri-generation system. Liu et al. (2004) investigated a hybrid tri-generation HVAC system in a multi-storey demonstration building, employing an ICE, liquid desiccant and vapour absorption technology. The investigation highlighted issues with conventional tri-generation systems that use a straight VAS, these include: (1) low grade waste heat from the prime mover is used directly to drive the cycle, leading to low energy performance and thus a COP_{th} of 0.7 or less, (2) how to match electrical and thermal loads of the building and lengthen the operating hours of the tri-generation system (thus improving benefits from the system), and (3) dehumidification in a VAS is achieved by cooling the air below its dew point. However, because the supply air requirement is 7–12 °C some re-heat is needed. This lowers the COP and can also cause health problems as the condensed water makes the coil surface a hot bed for bacteria. However, the authors found a way of avoiding the above problems when using a hybrid system consisting of liquid desiccant, VAS and thermal/desiccant storage. The findings are as follows: (1) the regeneration of liquid desiccant does not need high grade heat, only 70–80 °C, thus COP_{th} improvements of ~1.2–1.3 can be achieved (2) a thermal/desiccant storage tank can be used to match differences between thermal supply and demand, and (3) the liquid desiccant system will remove the latent loads, therefore avoiding the need to cool then re-heat. A simulation was carried out to analyse the performance of the building with the different system configurations i.e. with and without liquid desiccant. Results show that CO_2 emissions are reduced by about 40 % and energy storage plays an important role in the hybrid liquid desiccant system.

Qiu et al. (2012) have carried out an experimental investigation of a liquid desiccant system driven by the flue gases of a biomass boiler. The liquid desiccant system consists of two parts; first, the dehumidifier and secondly an

IEC. The desiccant dehumidifier used was of a direct contact counter flow falling film design, using a potassium formate ($CHKO_2$) solution. Internal cooling was provided from an evaporative device. The system was operated in autumn in Nottingham, UK and was found to be able decrease air temperature by 4 °C and relative humidity by 13 %. A total cooling capacity of the system was found to be 2.381 kW, the evaporative cooling device alone had a cooling capacity of 1.049 kW, illustrating the dehumidifier plays a significant role in supply air conditioning by reducing the airs enthalpy. Electrical consumption for fans and pumps totalled ~200 W; therefore the desiccant cooling systems COP_{el} was in the range of 11–12, an encouraging value. It was found that the heat from the flu gas was insufficient to heat the desiccant solution to the required 40 °C, only 30 °C was reached, as a result future work will look at improving the flue gas heat exchanger, leading to improved performance from the liquid desiccant system. This work has proved that desiccant systems are effective cooling devices in waste heat scenarios, however issues such as waste heat recovery effectiveness need to be addressed in order to maximise operating performance. The experimental results confirm that cooling is a result of lowering both humidity and air temperature.

Jradi and Riffat (2014c) present a novel biomass-fuelled micro-scale tri-generation system with an organic Rankine cycle and liquid desiccant cooling unit. Potassium formate is used as the desiccant working fluid. Results from laboratory experimental testing show the system can provide 9.6 kW heating, 6.5 kW cooling and 500 W electric power. The dehumidification-cooling system attains a COP_{th} of 0.86 and COP_{el} of 7.7. The overall efficiency of the micro-scale tri-generation system is 85 %. The high COP results demonstrate the potential of using liquid desiccant air conditioning technology in the development of efficient and effective tri-generation systems, and as an alternative to conventional separate energy conversion systems.

Badami and Portoraro (2009) have carried out a performance analysis of a tri-generation plant based on manufacturer data of a gas fired ICE and liquid desiccant air conditioning system to supply cooling (210 kW), heat (220 kW) and power (126 kW) to a teaching building at the Polytechnic University in Torino, Italy. Both an energetic and economic analysis was completed. The energetic analysis showed that the system could provide an 18 % energy saving in winter (no cooling) compared to a grid electricity and separate boiler heat production system. For the summer season, when the liquid desiccant system is operated, a total primary energy saving of −12 % was reported, showing the system has a higher primary energy consumption than that of a traditional separate production system. Therefore the convenience to operate the system during the summer months was based on its economic profitability. Economically, the system was assessed to have a pay-back period of 6.8–7.6 years, have an internal rate of return of 9.7–11.5 % and a net present value after 15 years of 200–220 k€. Variation in these assessments is due to the consideration of various national incentives for CHP and tri-generation systems. The performance analysis has shown that the adoption of tri-generation systems based on liquid desiccant technology is feasible economically, however energetically some questions do arise and should be considered before application. A limitation of the study is the assumption that all energy outputs from the system are consumed, thus all energy is fully exploited. Further analysis including time varying energy supply and demand may cause a poorer examination of the system.

Badami et al. (2012) have also carried out an analysis based on manufacturer data, comparing a tri-generation plant composed ICE and liquid desiccant with a micro gas turbine (MGT) and VAS. Energetic analysis carried out according to EU Directive 2004/8/EC (EU 2004) has shown that primary energy savings (PES) for the ICE/liquid desiccant system is greater than the MGT/VAS; 16.8 % compared to 9.1 % respectively. However further analysis using the more suitable tri-generation primary energy savings (TPES) methodology proposed in the literature (Chicco and Mancarella 2007), showed lower savings of 12.4 and 4.1 % respectively.

Section 2.5.4 has presented a review of the literature surrounding liquid desiccant air conditioning systems adopted in tri-generation system applications. Liquid desiccant based tri-generation systems can offer environmental, economic and operational advantages over other heat driven cooling cycles. VAS are rarely used in tri-generation applications of less than 10 kW (Pietruschka et al. 2006), therefore liquid desiccant is a particularly well suited option for the development of a tri-generation system at a domestic building level.

It is clear that the use of liquid desiccant air conditioning in tri-generation system applications is feasible, and has been demonstrated across a range of building scales. Some clear benefits liquid desiccant air conditioning can bring to a tri-generation system application include:

- Effective use of low grade thermal energy; high COP_{th} values reported.
- Removing the need to cool below the dew point to carry out dehumidification.

Careful consideration needs to be made so that the desiccant solution regeneration temperature can be achieved, this is critical for two reasons:

- A lower desiccant concentration will result in a reduced desiccant air conditioning system performance.
- Additional heating will reduce the potential savings of the system, thus impacting its feasibility as a replacement to a conventional separated system.

It is apparent that effective installation, optimisation and operation are all required in order to maximise the benefits attainable from the adoption of a liquid desiccant based tri-generation system. Otherwise issues such as an increase in energy demand seen in the work of Badami and Portoraro (2009) can result. Furthermore, the size and complexity of the liquid desiccant systems needs careful consideration, especially when integrated in a domestic application with other potentially large and complex equipment i.e. a fuel cell.

Next, Sect. 2.6 reviews the assessment methods used to evaluate the performance of tri-generation systems.

2.6 Assessment Methods for Tri-generation Systems

The literature reviewed has highlighted various methods and performance metrics used to assess the performance of tri-generation systems, particularly from an energetic and economic perspective. From the literature two clear approaches have been identified; (1) instantaneous performance and (2) seasonal performance.

Instantaneous performance looks predominately at maximum cooling output, cooling unit COP_{th} and tri-generation efficiency. Pilatowsky et al. (2007) assessed the cooling capacity at maximum power output from a PEMFC, whereas Qiu et al. (2012) looked at cooling output and COP_{th} based on specific inlet environmental conditions. This method is useful when looking at peak performance, but less useful when comparing it to other systems to determine expected savings, particularly over a specified time period.

Seasonal performance assessment of a tri-generation system investigates the energy savings over a specified time period. Badami and Portoraro (2009) and Badami et al. (2012) have carried out extensive work on the comparison of tri-generation systems from an energetic and economic perspective. The authors offer two methods of assessment (1) the commonly accepted methodology proposed by the EU—the Primary Energy Savings (PES) indicator (EU 2004) and (2) Tri-generation Primary Energy Savings (TPES), a method reported originally by Chicco and Mancarella (2007) to assess the energetic convenience offered by a tri-generation plant. Badami et al. (2012) feel TPES is a more suitable assessment method. The PES indicator does not include the performance of the cooling system; the heat employed to produce a cooling effect is considered as useful thermal energy. However, the TPES assessment method includes the performance of the cooling system, as a result the TPES method calculates lower, but more accurate savings compared to the PES method. Both methods rely on the performance data of the system being assessed against a reference comparison system.

Minciuc et al. (2003) have developed a more comprehensive method to assess the performance of tri-generation systems, approaching the analysis from the point of view of fuel savings and environmental impact. Minciuc et al. (2003) state that there are few publications which propose analysis methods for fuel savings and level of CO_2 emissions for tri-generation systems. The work has focussed on an ICE coupled to a VAS, however the approach and lessons learnt can be applied to any tri-generation system. Key findings include; relative fuel savings increase with increasing heat production, with maximum savings achieved when all heat produced in co-generation mode is used and no peak heat production method is required (boiler). Fuel savings increase with an increase in the COP of the cooling unit, CHP efficiency and heat recovery ratio. Finally, for the highest tri-generation system efficiency, the plant has to produce more electricity than heat or coolth.

Section 2.6 has presented some commonly adopted methods used to evaluate the performance of tri-generation systems. These methods will be fundamental to the theoretical and experimental evaluation of the tri-generation system presented in Chaps. 4 and 7 respectively. Next, Sect. 2.7 provides the conclusions to the literature review, with specific consideration as to their impact on the development of the proposed SOFC liquid desiccant tri-generation system.

2.7 Literature Review Conclusions and Impact on Tri-generation System Development

An extensive and thorough review of literature has been presented, covering five key topics: fuel cell technology in Sect. 2.2, desiccant air conditioning in Sect. 2.3, combined heat and power in Sect. 2.4, tri-generation systems in Sect. 2.5 and tri-generation assessment methods in Sect. 2.6.

The aim of the review has been to (1) highlight the current gap regarding a SOFC liquid desiccant tri-generation system, (2) demonstrate the feasibility of the development of such a system, and (3) provide a comprehensive background study on the state of the art of the technology. All literature has been selected based on its appropriateness and suitability in addressing the central aim of the thesis; the design, development and testing of an efficient and effective SOFC liquid desiccant tri-generation system. The main literature review conclusions are summarised as follows.

- No work has been reported on an SOFC liquid desiccant tri-generation system. Literature found regarding SOFC tri-generation is predominately simulation or larger commercial applications utilising VAS, with little work on experimental based domestic building studies. Furthermore, there is little work reported on fuel cell tri-generation in domestic built environment applications. Work that has been completed is again simulation based. The current highlighted gaps are believed to be a result of three key factors (1) cost, (2) system size/complexity and (3) relative infancy of fuel cell/heat driven cooling technology, particularly at a domestic scale.
- The low electrical requirement and high COP_{th} makes stand-alone evaporative based liquid desiccant systems an attractive option for effective waste heat utilisation in tri-generation system applications. The use of membrane based contactors eliminates the issues of desiccant carry over. Potassium formate used as a working fluid demonstrates good dehumidification capacity, particularly in high humidity applications. There is limited published and available literature within the public domain regarding the specific physical characterisation of the potassium formate solution in the required operating range for liquid desiccant air conditioning applications.
- Opportunities that the literature review have highlighted are; the high operating electrical efficiency and stable operation of SOFC technology, coupled with the effective use of low-grade thermal energy seen in stand-alone liquid desiccant air conditioning demonstrates the potential for the first of its kind SOFC liquid desiccant tri-generation system. The outlet water temperature in a SOFC

CHP system (60 °C) is in good thermal agreement with the required desiccant solution regeneration temperatures (45–50 °C). Confirming the feasibility of integrating the two technologies into an efficient and effective tri-generation system.
- Challenges that the literature review has highlighted are; system integration, optimisation and energy utilisation is not an issue in theoretical or laboratory based tri-generation systems, thus high system efficiencies have been reported. However, in real world building applications, effective energy utilisation and low COP_{th} in the heat driven cooling cycle can have a marked impact on tri-generation system performance and thus feasibility.

Further to the general literature review conclusions provided above. Specific conclusions relevant to the design, development and testing of the novel system are presented below.

- Technical constraints will mean that the SOFC will run continuously. The balance between energy supply and energy demand needs careful and thoughtful consideration to ensure high system performance. To ensure high tri-generation system performance, there needs to be a constant electrical and thermal energy demand, or an effective way of storing it. Electricity can be imported/exported to the grid. However, the issue of heat and coolth will be more problematic. This issue will be even greater outside the hours of building inhabitation and in spring/autumn when air conditioning will not be required.
- Due to the issues other researches have encountered whilst coupling off the shelf components into a tri-generation system, careful consideration as to the thermal pairing of the SOFC CHP and liquid desiccant systems is required.
- The developed tri-generation system should have the option to be modified once built in order to facilitate any changes required to improve its performance i.e. inclusion of thermal energy storage, heat exchangers etc.

The literature review has highlighted that there is a clear need for research in the field of SOFC liquid desiccant tri-generation and domestic scale fuel cell tri-generation systems. This is due to the patent operational advantages the system can deliver in terms of energy efficiency, operating costs and energy security. This thesis aims to address the highlighted research gaps through theoretical and experimental means.

References

Abdel-Salam, A. H., G. Ge and C. J. Simonson. 2013. Performance analysis of a membrane liquid desiccant air-conditioning system. *Energy and Buildings* 62: 559–569.

Al-Sulaiman, F.A., I. Dincer, and F. Hamdullahpur. 2010. Energy analysis of a trigeneration plant based on solid oxide fuel cell and organic Rankine cycle. *International Journal of Hydrogen Energy* 35(10): 5104–5113.

Al-Sulaiman, F.A., F. Hamdullahpur, and I. Dincer. 2011. Trigeneration: A comprehensive review based on prime movers. *International Journal of Energy Research* 35(3): 233–258.

ASHRAE. 2009. *2009 ASHRAE Handbook—Fundamentals (I-P Edition)*. Atlanta, USA, American Society of Heating, Refrigerating and Air-Conditioning Engineers, Inc.

Badami, M., and A. Portoraro. 2009. Performance analysis of an innovative small-scale trigeneration plant with liquid desiccant cooling system. *Energy and Buildings* 41(11): 1195–1204.

Badami, M., A. Portoraro, and G. Ruscica. 2012. Analysis of trigeneration plants: Engine with liquid desiccant cooling and micro gas turbine with absorption chiller. *International Journal of Energy Research* 36(5): 579–589.

Beausoleil-Morrison, I., and K. Lombardi. 2009. The calibration of a model for simulating the thermal and electrical performance of a 2.8 kWAC solid-oxide fuel cell micro-cogeneration device. *Journal of Power Sources* 186(1): 67–79.

Beaussoleil-Morrison, I. 2008. *An experimental and simulation-based investigation of the performance of small scale fuel cell and combustion-based cogeneration devices serving residential buildings*. Annex 42 of the International Energy Agency's Energy Conservation in Buildings and Community Systems Programme.

Bergero, S., and A. Chiari. 2010. Performance analysis of a liquid desiccant and membrane contactor hybrid air-conditioning system. *Energy and Buildings* 42(11): 1976–1986.

Bhatti, M. S., J. F. O'Brien, I. Reyzin, M. Grieve and S. M. Kelly. 2010. *Solid oxide fuel cell assisted air conditioning system*. USA.

Bianchi, M., A. De Pascale, and P.R. Spina. 2012. Guidelines for residential micro-CHP systems design. *Applied Energy* 97: 673–685.

Blomen, L. J., M.J. Blomen, M.N. Mugerwa. 1993. *Fuel Cell Systems*. New York.

Bonke, M. 2007. Novel liquid desiccant dehumidification system (DEHUMID). Co-operative Research Projects, COOP-CT-2005-016957

Boyd, J. 2008. *IEEE Spectrum: Home fuel cells to sell in Japan*. Retrieved 27/04/2012, from http://spectrum.ieee.org/energy/renewables/home-fuel-cells-to-sell-in-japan.

Bradley, A. 2013. *Technology shift in micro-CHP: Fuel cell outsells engines for the first time*. News-releases Retrieved 03/02/2013, 2013, from http://www.prnewswire.com/news-releases/technology-shift-in-micro-chp-fuel-cell-outsells-engines-for-the-first-time-214928111.html.

Brouwer, J. 2010. On the role of fuel cells and hydrogen in a more sustainable and renewable energy future. *Current Applied Physics* 10(2, Suppl.): S9–S17.

Buker, M.S., and S.B. Riffat. 2015. Recent developments in solar assisted liquid desiccant evaporative cooling technology—A review. *Energy and Buildings* 96: 95–108.

Chicco, G., and P. Mancarella. 2007. Trigeneration primary energy saving evaluation for energy planning and policy development. *Energy Policy* 35(12): 6132–6144.

Choudhury, A., H. Chandra, and A. Arora. 2013. Application of solid oxide fuel cell technology for power generation—A review. *Renewable and Sustainable Energy Reviews* 20: 430–442.

Conde, M. 2007. Liquid desiccant-based air-conditioning systems—LDACS. in *1st European Conference on Polygeneration,* Tarragona, Spain.

Dai, Y.J., R.Z. Wang, H.F. Zhang, and J.D. Yu. 2001. Use of liquid desiccant cooling to improve the performance of vapor compression air conditioning. *Applied Thermal Engineering* 21(12): 1185–1202.

Daou, K., R.Z. Wang, and Z.Z. Xia. 2006. Desiccant cooling air conditioning: A review. *Renewable and Sustainable Energy Reviews* 10(2): 55–77.

Darwish, M.A. 2007. Building air conditioning system using fuel cell: Case study for Kuwait. *Applied Thermal Engineering* 27(17–18): 2869–2876.

Das, R.S., P.K. Saha, and S. Jain. 2012. *Investigations on solar energy driven liquid desiccant cooling systems for tropical climates*. Melbourne: Australian Solar Energy Society. Australian Solar Council.

Deng, J., R.Z. Wang, and G.Y. Han. 2011. A review of thermally activated cooling technologies for combined cooling, heating and power systems. *Progress in Energy and Combustion Science* 37(2): 172–203.

Duan, Z., Zhan, Changhong., Zhang, Xingxing., Mustafa, Mahmud., Zhao, Xudong., Alimohammadisagvand, Behrang., Hasan, Ala (2012). Indirect evaporative cooling: Past, present and future potentials. *Renewable and Sustainable Energy Reviews* 16(9): 6823–6850.

Dwyer, T. 2014. Liquid desiccants for dehumidification in building air conditioning systems. *The CIBSE Journal CPD Programme*.

El-Gohary, M.M. 2013. Economical analysis of combined fuel cell generators and absorption chillers. *Alexandria Engineering Journal* 52(2): 151–158.

ElianEnergy. 2011. *Fuel cell cars*. Retrieved 09/05/2012, from http://www.future-alternative-energy.net/fuel-cell-cars.html.

Ellamla, H.R., I. Staffell, P. Bujlo, B.G. Pollet, and S. Pasupathi. 2015. Current status of fuel cell based combined heat and power systems for residential sector. *Journal of Power Sources* 293: 312–328.

Elmer, T. and S. B. Riffat. 2014. State of the art review: Fuel cell technologies in the domestic built environment. in *Progress in Sustainable Energy Technologies Vol II: Creating Sustainable Development*, ed. by I. Dincer, A. Midilli and H. Kucuk. Cham. New York: Springer International Publishing, p. 247–271.

Elmer, T., M. Worall, S. Wu, and S.B. Riffat. 2015. Fuel cell technology for domestic built environment applications: State of-the-art review. *Renewable and Sustainable Energy Reviews* 42: 913–931.

EU (2004) Directive 2004/8/EC of the European Parliament And Of The Council—On the promotion of cogeneration based on a useful heat demand in the internal energy market and amending Directive 92/42/EEC. *Official Journal of the European Union* 11.

Farhad, S., F. Hamdullahpur, and Y. Yoo. 2010. Performance evaluation of different configurations of biogas-fuelled SOFC micro-CHP systems for residential applications. *International Journal of Hydrogen Energy* 35(8): 3758–3768.

Fan, Liangdong, Chengyang Wang, Mingming Chen, and Bin Zhu. 2013. Recent development of ceria-based (nano) composite materials for low temperature ceramic fuel cells and electrolyte-free fuel cells. *Journal of Power Sources* 234: 154–174.

FCT. 2013. *Fuel cell today: The leading authority on fuel cells: The Fuel Cell Today Industry Review 2013*. http://www.fuelcelltoday.com/media/1889744/fct_review_2013.pdf.

Fong, K.F., and C.K. Lee. 2014. Investigation on zero grid-electricity design strategies of solid oxide fuel cell trigeneration system for high-rise building in hot and humid climate. *Applied Energy* 114: 426–433.

Fubara, T.C., F. Cecelja, and A. Yang. 2014. Modelling and selection of micro-CHP systems for domestic energy supply: The dimension of network-wide primary energy consumption. *Applied Energy* 114: 327–334.

Gencoglu, M.T., and Z. Ural. 2009. Design of a PEM fuel cell system for residential application. *International Journal of Hydrogen Energy* 34(12): 5242–5248.

Gigliucci, G., L. Petruzzi, E. Cerelli, A. Garzisi, and A. La Mendola. 2004. Demonstration of a residential CHP system based on PEM fuel cells. *Journal of Power Sources* 131(1–2): 62–68.

Gommed, K., and G. Grossman. 2007. Experimental investigation of a liquid desiccant system for solar cooling and dehumidification. *Solar Energy* 81(1): 131–138.

Harrison, J. 2012. E.ON—Smart Homes with Fuel Cell micro CHP. *Smart Hydrogen & Fuel Cell Power Conference*, Birmingham.

Havelský, V. 1999. Energetic efficiency of cogeneration systems for combined heat, cold and power production. *International Journal of Refrigeration* 22(6): 479–485.

Hawkes, A., G. Tiravanti, and M. Leach. 2005. Routes to energy efficiency: Complementary energy service products in the UK residential sector. *European Council for an Energy Efficient Economy (ECEEE)* 2: 531–540.

Hawkes, A., I. Staffell, D. Brett, and N. Brandon. 2009. Fuel cells for micro-combined heat and power generation. *Energy & Environmental Science* 2(7): 729–744.

Huangfu, Y., J.Y. Wu, R.Z. Wang, X.Q. Kong, and B.H. Wei. 2007. Evaluation and analysis of novel micro-scale combined cooling, heating and power (MCCHP) system. *Energy Conversion and Management* 48(5): 1703–1709.

Jain, S., and P.K. Bansal. 2007. Performance analysis of liquid desiccant dehumidification systems. *International Journal of Refrigeration* 30(5): 861–872.

Jradi, M. and S. Riffat. 2014a. Energy performance of an innovative liquid desiccant dehumidification system with a counter-flow heat and mass exchanger using potassium formate. *Renewable Bioresources* 2(5).

Jradi, M., and S. Riffat. 2014b. Experimental investigation of a biomass-fuelled micro-scale trigeneration system with an organic Rankine cycle and liquid desiccant cooling unit. *Energy* 71: 80–93.

Kazempoor, P., V. Dorer, and A. Weber. 2011. Modelling and evaluation of building integrated SOFC systems. *International Journal of Hydrogen Energy* 36(20): 13241–13249.

Khatri, K.K., D. Sharma, S.L. Soni, and D. Tanwar. 2010. Experimental investigation of CI engine operated micro-trigeneration system. *Applied Thermal Engineering* 30(11–12): 1505–1509.

Kong, X.Q., R.Z. Wang, J.Y. Wu, X.H. Huang, Y. Huangfu, D.W. Wu, and Y.X. Xu. 2005. Experimental investigation of a micro-combined cooling, heating and power system driven by a gas engine. *International Journal of Refrigeration* 28(7): 977–987.

Kozubal, W., J. Woods, J. Burch, A. Boranian and T. Merrigan. 2011. *Desiccant enhanced evaporative air-conditioning (DEVap): Evaluation of a new concept in ultra efficient air conditioning*. Technical Report NREL/TP-5500-49722, January 2011, Contract No. DE-AC36-08GO28308.

Kozubal, E., J. Woods and R. Judkoff. 2012. *Development and analysis of desiccant enhanced evaporative air conditioner prototype*. National Renewable Energy Laboratory.

Kuhn, V., J. Klemeš, and I. Bulatov. 2008. MicroCHP: Overview of selected technologies, products and field test results. *Applied Thermal Engineering* 28(16): 2039–2048.

Liu, S. 2008. *A novel heat recovery/desiccant cooling system*. PhD: The University of Nottingham.

Liu, X.H., K.C. Geng, B.R. Lin, and Y. Jiang. 2004. Combined cogeneration and liquid-desiccant system applied in a demonstration building. *Energy and Buildings* 36(9): 945–953.

Lowenstein, A. 2008. Review of liquid desiccant technology for HVAC applications. *American Society of Heating, Refrigerating and Air-Conditioning Engineers* 14(6): 819–839.

Lychnos, G., and P.A. Davies. 2012. Modelling and experimental verification of a solar-powered liquid desiccant cooling system for greenhouse food production in hot climates. *Energy* 40(1): 116–130.

Maghanki, M.M., B. Ghobadian, G. Najafi, and R.J. Galogah. 2013. Micro combined heat and power (MCHP) technologies and applications. *Renewable and Sustainable Energy Reviews* 28: 510–524.

Malico, I., A.P. Carvalhinho, and J. Tenreiro. 2009. Design of a trigeneration system using a high-temperature fuel cell. *International Journal of Energy Research* 33(2): 144–151.

Margalef, P., and S. Samuelsen. 2010. Integration of a molten carbonate fuel cell with a direct exhaust absorption chiller. *Journal of Power Sources* 195(17): 5674–5685.

Mei, L., and Y.J. Dai. 2008. A technical review on use of liquid-desiccant dehumidification for air-conditioning application. *Renewable and Sustainable Energy Reviews* 12(3): 662–689.

Míguez, J. L., S. P. Murillo, J., and L. M. López. 2004. Feasibility of a new domestic CHP trigeneration with heat pump: I. Design and development. *Applied Thermal Engineering* 24(10): 1409–1419.

Minciuc, E., O. Le Corre, V. Athanasovici, and M. Tazerout. 2003. Fuel savings and CO_2 emissions for tri-generation systems. *Applied Thermal Engineering* 23(11): 1333–1346.

Minh, N.Q. 2004. Solid oxide fuel cell technology—features and applications. *Solid State Ionics* 174(1–4): 271–277.

O'Hayre, R., S. K. Cha, W. Colella and F. B. Prinz. 2006. *Fuel cell fundamentals*. New York: Wiley.

Ouazia, B., H. Barhoun, K. Haddad, M. Armstrong, R. G. Marchand and F. Szadkowski. 2009. Desiccant-evaporative cooling system for residential buildings. in *12th Canadian Conference on Building Science and Technology*, Montréal, Québec, Institute for Research in Construction.

Peht, M., M. Cames, C. Fischer, B. Prateorius, L. Schneider, K. Schumacher, and J. Voss. 2006. *Micro cogeneration towards dencentralized energy systems*. Berlin: Springer.

Pérez-Lombard, L., J. Ortiz, and C. Pout. 2008. A review on buildings energy consumption information. *Energy and Buildings* 40(3): 394–398.

Pietruschka, D., U. Eicker, M. Huber, and J. Schumacher. 2006. Experimental performance analysis and modelling of liquid desiccant cooling systems for air conditioning in residential buildings. *International Journal of Refrigeration* 29(1): 110–124.

Pilatowsky, I., R.J. Romero, C.A. Isaza, S.A. Gamboa, W. Rivera, P.J. Sebastian, and J. Moreira. 2007. Simulation of an air conditioning absorption refrigeration system in a co-generation process combining a proton exchange membrane fuel cell. *International Journal of Hydrogen Energy* 32(15): 3174–3182.

Pilatowsky, I., R.J. Romero, C.A. Isaza, S.A. Gamboa, P.J. Sebastian, and W. Rivera. 2011. *Cogeneration fuel cell—Sorption air conditioning systems*. London: Springer.

Porteiro, J., J. L. Míguez, S. Murillo, L. M. López. 2004. Feasibility of a new domestic CHP trigeneration withheat pump: II. Availability analysis. *Applied Thermal Engineering* 24(10): 1421–1429.

Qiu, G., H. Liu and S. Riffat. 2012. Experimental investigation of a liquid desiccant cooling system driven by flue gas waste heat of a biomass boiler. *International Journal of Low-Carbon Technologies* 8: 165–172.

Qiu, G.Q., and S.B. Riffat. 2010. Experimental investigation on a novel air dehumidifier using liquid desiccant. *International Journal of Green Energy* 7(2): 174–180.

Ren, H., and W. Gao. 2010. Economic and environmental evaluation of micro CHP systems with different operating modes for residential buildings in Japan. *Energy and Buildings* 42(6): 853–861.

Riffat, S. 2012. *EU project proposal: Durable Solid Oxide Fuel Cell Tri-generation System for Low Carbon Buildings*. Project 303454 - TriSOFC.

Sammes, N.M., and R. Boersma. 2000. Small-scale fuel cells for residential applications. *Journal of Power Sources* 86(1–2): 98–110.

Smith, S.T., V.I. Hanby, and C. Harpham. 2011. A probabilistic analysis of the future potential of evaporative cooling systems in a temperate climate. *Energy and Buildings* 43(2–3): 507–516.

Sonar, D., S. L. Soni and D. Sharma. 2014. Micro-trigeneration for energy sustainability: Technologies, tools and trends. *Applied Thermal Engineering* 71: 790–796.

Srikhirin, P., S. Aphornratana, and S. Chungpaibulpatana. 2001. A review of absorption refrigeration technologies. *Renewable and Sustainable Energy Reviews* 5(4): 343–372.

Staffell, I. (2009). Fuel Cells for Domestic Heat and Power: Are They Worth It?, The University of Birmingham.

Staffell, I., and R. Green. 2013. The cost of domestic fuel cell micro-CHP systems. *International Journal of Hydrogen Energy* 38(2): 1088–1102.

Steele, B.C.H. 1999. Fuel-cell technology: Running on natural gas. *Nature* 400(6745): 619–621.

Steinberger-Wilckens, R. 2013. Fuel cells and hydrogen joint undertaking: Evaluating the performance of fuel cells in european energy supply grids (FC-EuroGrid). *Fuel Cells and Hydrogen Joint Undertaking*.

Studak, J. and J. L. Peterson. 1988. *A preliminary evaluation of alternative liquid desiccants for a hybrid desiccant air conditioner*. Energy Systems Laboratory. http://esl.tamu.edu.

Vourliotakis, G., D. Giannopoulos and M. Founti. 2010. *Potentials of fuel cells as μ-CHP systems for domestic applications in the framework of energy efficient and sustainable districts*. FC-DISTRICT project, 7th FP, CP-IP 260105 FC-DISTRICT.

Wang, J., C. M. Li, J. H. Liu, S. C. Liu and J. Chen. 2009 *A new air-conditioning system of liquid desiccant and evaporation cooling*. College of Urban Construction & Environmental Engineering University of Shanghai forScience and Technology Shanghai, P.R. China.

Welch, T. 2008. *CIBSE knowledge series: KS13—Refrigeration*. H. Carwarardine and K. Butcher. London: CIBSE Publications.

Woods, J., and E. Kozubal. 2013. A desiccant-enhanced evaporative air conditioner: Numerical model and experiments. *Energy Conversion and Management* 65: 208–220.

Wu, D.W., and R.Z. Wang. 2006. Combined cooling, heating and power: A review. *Progress in Energy and Combustion Science* 32(5–6): 459–495.

Wu, J.Y., J.L. Wang, S. Li, and R.Z. Wang. 2014. Experimental and simulative investigation of a micro-CCHP (micro combined cooling, heating and power) system with thermal management controller. *Energy* 68: 444–453.

Yadav, Y.K., and S.C. Kaushik. 1991. Psychometric techno-economics assessment and parametric study of vaporcompression and solid/liquid hybrid air-conditioning system. *Heat Recovery CHP* 11(6): 563–572.

Yu, Z., Jitian Han, and Xianqi Cao. 2011. Investigation on performance of an integrated solid oxide fuel cell and absorption chiller tri-generation system. *International Journal of Hydrogen Energy* 36(19): 12561–12573.

Zhang, L.Z. 2006a. Energy performance of independent air dehumidification systems with energy recovery measures. *Energy* 31(8–9): 1228–1242.

Zhang, L.Z. 2006b. Mass diffusion in a hydrophobic membrane humidification/dehumidification process: The effects of membrane characteristics. *Separation Science and Technology* 41(8): 1565–1582.

Zink, F., Yixin Lu, and Laura Schaefer. 2007. A solid oxide fuel cell system for buildings. *Energy Conversion and Management* 48(3): 809–818.

Chapter 3
Liquid Desiccant Air Conditioning System Modelling and Analysis

3.1 Introduction

This chapter provides a theoretical performance investigation of the liquid desiccant air conditioning technology that will be integrated alongside the SOFC in the tri-generation system. The aims of this chapter are to (1) assess a variety of liquid desiccant solutions with regard to their potential for dehumidification and suitability for integration into a liquid desiccant based tri-generation system, and (2) develop validated mathematical models to evaluate the dehumidifier and regenerator heat and mass exchangers. The evaluation of desiccant working fluids and the development of mathematical models will assist the fuel cell tri-generation system in the following respects:

(1) Justification of the selection of an appropriate working fluid for the liquid desiccant air conditioning system.
(2) Improved understating of the performance and operating process of liquid desiccant air conditioning systems under a variety of environmental and operating conditions. This evaluation will support the experimental work presented in Chaps. 5 and 6.
(3) Selection of optimised conditions to operate a liquid desiccant air conditioning system in a tri-generation system set-up, with specific consideration for SOFC integration.
(4) To confirm the feasibility of integrating liquid desiccant air conditioning and SOFC technology to form a reliable and efficient tri-generation system.
(5) Enable the development of a SOFC tri-generation system model, presented in Chap. 4.

The modelling work presented in this chapter does not aim to progress the state of the art in the theoretical analysis of liquid desiccant air conditioning systems, but to facilitate a better understanding of the process and performance of such

T. Elmer, *A Novel SOFC Tri-generation System for Building Applications*,
Springer Theses, DOI 10.1007/978-3-319-46966-9_3

systems. Particularly so that conclusions may be drawn that are relevant to the integration of the liquid desiccant air conditioning system with a SOFC in a tri-generation system set-up. The work in Sect. 3.2 provides a contribution to knowledge through the assimilation of various sources of literature in order to obtain the necessary data, over the required operating range, to successfully evaluate (theoretically and experimentally) a liquid desiccant air conditioning system operating with a potassium formate solution. This data has not been previously published and available within the public domain.

This chapter is split into five sections: Sect. 3.2 evaluates three desiccant solutions with regards to their dehumidification capacity, performance and suitability for use in a tri-generation system. Section 3.3 introduces the mathematical models used to evaluate the dehumidifier and regenerator membrane heat and mass exchangers. Section 3.4 validates the developed mathematical model with published experimental data. Section 3.5 presents a dehumidifier and regenerator parametric analysis. Section 3.6 provides the chapter's conclusions, summarising the results from the parametric analysis, and the implications these results have on the development of the fuel cell tri-generation system.

3.1.1 Desiccant System Description

A liquid desiccant air conditioning system consists of two main components (1) the desiccant dehumidifier, and (2) the regenerator/regeneration heat source, as shown in Fig. 3.1. The role of the dehumidifier is to reduce the moisture content of the building supply air to provide a comfortable environment for occupants. As moisture is absorbed by the liquid desiccant solution, it becomes dilute and thus its ability to absorb moisture is reduced. In order to re-use the desiccant solution, the regenerator and regenerator heat source are used to evaporate off the moisture gained, thus re-concentrating the liquid desiccant solution. Following

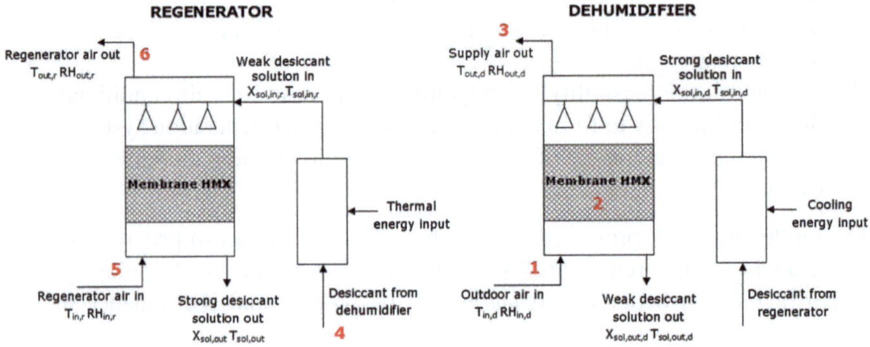

Fig. 3.1 Liquid desiccant system layout and operation

regeneration, the liquid desiccant solution should be reduced in temperature close to that of the supply air requirement, to facilitate both latent and sensible cooling.

In this thesis, a membrane based dehumidifier and regenerator heat and mass exchanger (HMX) have been employed to overcome issues of desiccant entrainment in the supply airstream. Because the dehumidification and regeneration processes involve both the transfer of heat and mass (water vapour), they are referred to as heat and mass exchangers. Next, Sect. 3.2 provides an evaluation of a suitable liquid desiccant solution for the SOFC tri-generation system.

3.2 Desiccant Solution Evaluation

Liquid desiccant solutions used for air conditioning applications should, according to Jain and Bansal (2007) possess a range of physical characteristics; low vapour pressure, low density and viscosity, good heat transfer characteristics and favourable surface tension as this influences the wetting of contacting media. There are a variety of liquid desiccant solutions available for air conditioning applications; they can be split into three broad groups: halide salts, glycols and weak salt solutions. Application of a desiccant material is dependent upon cost, operation and source of thermal energy (Enteria and Mizutani 2011). In this section three desiccant solutions are evaluated with regards to their potential for dehumidification and suitability for application in a SOFC tri-generation system. The solutions considered are:

(1) Lithium chloride (LiCl)
(2) Calcium chloride ($CaCl_2$)
(3) Potassium formate ($CHKO_2$)

LiCl and $CaCl_2$, known as halide salts, have been widely used as working fluids in liquid desiccant air conditioning systems. The advantages of these materials are that they are strong desiccants; LiCl can dry air to 11 % RH (Lowenstein 2008). However, the halide salts are extremely corrosive and cause significant damage to air conditioning equipment (heat exchangers, pipes etc.). Titanium is one of the few materials that can be used, however it is very expensive. In response to the shortcomings of the halide salt desiccant solutions, other options have been explored. Salts of weak organic acids such as potassium or sodium formate have been used. These solutions have low toxicity, density and viscosity, are less corrosive, and they can, at the correct concentration achieve sufficient dehumidification potential. The concentrations of a $CHKO_2$ solution for an air conditioning application needs to be greater than that of halide salts, for example, in terms of dehumidification potential, $CHKO_2$ at a 50 % concentration by mass is roughly equivalent to a 27 % LiCl solution by mass. Although it is a weaker desiccant than the halide salts, for air conditioning applications its ability to dehumidify air below 30 % RH makes it an attractive desiccant solution (Lowenstein 2008).

As discussed in Sect. 3.1.1, dehumidification in liquid desiccant air conditioning systems is driven by a vapour pressure differential between the hot humid air (high vapour pressure) and the liquid desiccant solution (low vapour pressure). An assessment of the dehumidification potential of the LiCl, $CaCl_2$ and $CHKO_2$ desiccant solutions is achieved by comparing the vapour pressures of the different solutions over a range of solution concentrations and comparing the results. The vapour pressures of the LiCl and $CaCl_2$ solutions have been calculated using a routine based on extensive empirical data presented by Conde (2009). The vapour pressure of the $CHKO_2$ solution is determined using a routine, presented in Appendix A (A1.1), taken from the work of James (1998). The routines presented are valid in the 0–100 °C and 0–0.8 solution mass concentration range; thus covering the requirements of a typical air conditioning application. The three desiccant solutions have been evaluated at a set solution temperature of 20 °C, over a concentration range of 0.2–0.8. Figure 3.2 shows the vapour pressure comparison of the three solutions. Identification of the solution operating concentrations is based upon values which will provide a dehumidification potential across a range of summer air conditions. At an air dry bulb temperature of 20 °C, the ranges selected will dehumidify air from 65 % RH upwards at the lower solution concentrations, to 30 % RH upwards at the higher solution concentrations. The ranges selected also agree well with published literature and do not pose issues of crystallisation.

Table 3.1 summarises the typical operating concentrations LiCl, $CaCl_2$ and $CHKO_2$ are used at in liquid desiccant air conditioning systems. Other important considerations such as corrosiveness, cost, regeneration temperature and density are also detailed.

It is evident from the data presented in Fig. 3.2 and Table 3.1 that the LiCl and $CaCl_2$ solutions operate at lower concentrations comparted to the $CHKO_2$ solution. LiCl and $CaCl_2$ show, based on vapour pressure, better dehumidification

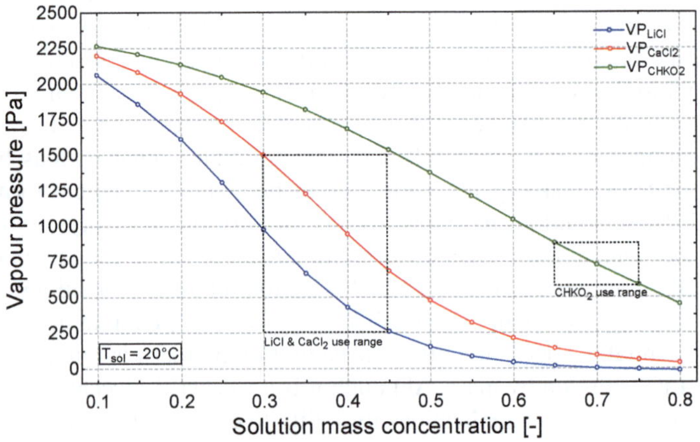

Fig. 3.2 Vapour pressure comparison for LiCl, $CaCl_2$ and $CHKO_2$

Table 3.1 Typical operating conditions of different desiccant solutions

Parameter	LiCl	$CaCl_2$	$CHKO_2$
Solution mass concentration	0.3–0.45	0.3–0.45	0.65–0.75
Vapour pressure (Pa)	974.4–257.4	1225–684.1	880.2–584.5
Corrosiveness	High	High	Low
Cost	High	High	Low
Regeneration temperature (°C)	70–90	70–90	50–70
Density (kg m^{-3})	2068	2150	1908

performance compared to $CHKO_2$. However, the $CHKO_2$ solution shows better environmental performance, is less harmful to human beings, and is cheaper. As a result, the desiccant air conditioning equipment can be cheaper as there is less need for expensive materials highly resistant to corrosion, also desiccant entrainment into supply airstreams poses less of a threat. Furthermore, the temperature at which the desiccant solution may be regenerated (water vapour driven off) is critical, particularly in a tri-generation system, as it needs to match the thermal output of the prime mover, in this case the SOFC. In general the halide salts (LiCl and $CaCl_2$) require higher regeneration temperatures in the range of 70–90 °C. As demonstrated in Sect. 3.5.2, the $CHKO_2$ solution may be regenerated anywhere between 45 and 70 °C. The $CHKO_2$ solutions regeneration temperature is therefore in better thermal agreement with the SOFC thermal output which is usually around 60 °C in a SOFC CHP building application (Pilatowsky et al. 2011; Jradi and Riffat 2014a, c).

Section 3.2 has evaluated three desiccant solutions that could be employed in the novel tri-generation system. Compared to $CHKO_2$, LiCl and $CaCl_2$ show better dehumidification potential at their typical operating concentrations. However, the $CHKO_2$ solution possesses many characteristics that make it favourable, particularly in a tri-generation system application. Positive attributes include low density and viscosity, a lower regeneration temperature requirement, negligible environmental impact, low corrosiveness and cost. As a result, a $CHKO_2$ solution is the preferred working fluid to be used in the SOFC tri-generation system. As a result, Sect. 3.2.1 will investigate the dehumidification performance of the $CHKO_2$ solution in more detail.

3.2.1 Potassium Formate Solution Climatic Performance Investigation

The aim of this section is to investigate the performance of the $CHKO_2$ solution, with particular regards to its potential for dehumidification under a range of operating and climatic conditions. Following this, the required desiccant operating concentration range is selected.

Figure 3.3 shows the lines of vapour pressure for the $CHKO_2$ desiccant solution over a range of temperatures and concentrations. It is evident as solution

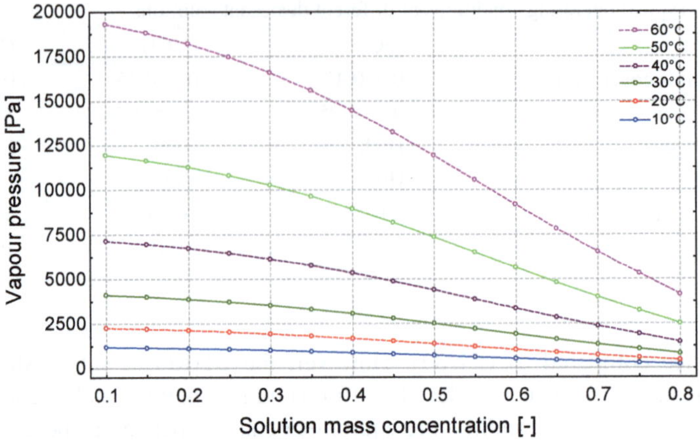

Fig. 3.3 $CHKO_2$ vapour pressures

concentration increases and temperature decreases, the vapour pressure of the solution reduces. A reduction in vapour pressure of the desiccant solution represents a greater dehumidification potential as it generates a greater vapour pressure differential between the hot humid ambient air and the desiccant solution. Figure 3.3 shows the $CHKO_2$ solution over a wide range of operating conditions. In an air conditioning application, it is likely the solution will be used in a much narrower range; 0.6–0.7 solution mass concentration at a temperature of 20–30 °C in the dehumidifier and 55 °C in the regenerator.

Figure 3.4 shows a chart of a range of climatic air conditions and the equivalent air moisture content under these conditions. Lines of the $CHKO_2$ equivalent absolute humidity (ω_{eq}), calculated using Eq. 3.27, are drawn over the ambient air conditions. For dehumidification to occur the equivalent moisture content of the desiccant solution needs to be less than that of the ambient air. A greater difference in vapour pressure demonstrates a greater potential for mass transfer (latent cooling). Furthermore, a solution temperature that is lower than that of the humid air will cause (sensible) cooling of the air.

Figure 3.4 alongside Table 3.2 shows the eight climatic conditions investigated, and the required desiccant concentration (temperature dependant) to perform dehumidification, under the assumption of perfect heat and mass transfer. At a solution temperature of 10 °C, the solution concentration needs to be greater than 0.5 for dehumidification to occur in all eight cases. At a solution temperature of 20 °C, the solution concentration needs to be greater than 0.725 for dehumidification to occur in all eight cases. At a solution temperature of 30 °C, the solution cannot dehumidify cases one and two, but will generate dehumidification in cases three to eight, with a solution concentration of 0.7.

Based on the assumption that the dehumidifier solution temperature will be in the 20–30 °C temperature range, Sect. 3.2.1 has established that a $CHKO_2$ solution will need to be used at a concentration of 0.65–0.7 for it to be an

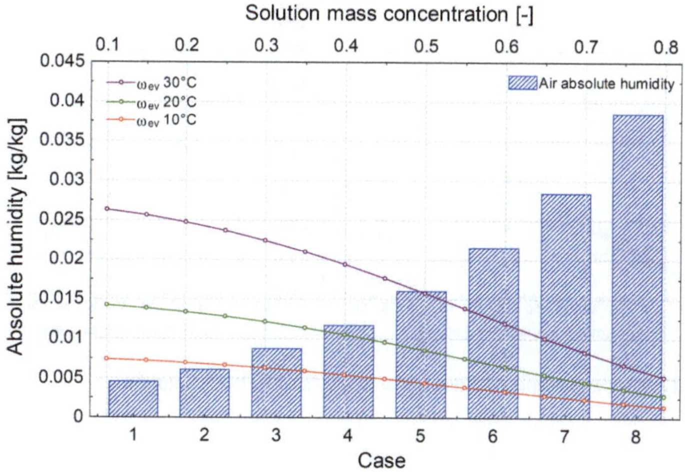

Fig. 3.4 Dehumidification potential of $CHKO_2$ in various climatic conditions

Table 3.2 Eight cases for air dehumidification

Case	T_a (°C)	RH_a	Description	Concentration to dehumidify		
				$T_{sol} = 10$ °C	$T_{sol} = 20$ °C	$T_{sol} = 30$ °C
1	10	0.6	Cold/dry	≥0.5	≥0.725	–
2	10	0.8	Cold/humid	≥0.325	≥0.65	–
3	20	0.6	Mild/dry	≥0.1	≥0.5	≥0.7
4	20	0.8	Mild/humid	≥0.1	≥0.325	≥0.625
5	30	0.6	Warm/dry	≥0.1	≥0.1	≥0.5
6	30	0.8	Warm/humid	≥0.1	≥0.1	≥0.325
7	40	0.6	Hot/dry	≥0.1	≥0.1	≥0.1
8	40	0.8	Hot/humid	≥0.1	≥0.1	≥0.1

effective dehumidifier across a range of summer climatic conditions. Furthermore, Sect. 3.5.2 demonstrates that the required solution temperature for regeneration in this concertation range is suitable for tri-generation system integration.

Next, Sect. 3.2.2 presents the thermophysical properties of the $CHKO_2$ solution.

3.2.2 Potassium Formate Thermophysical Characterisation

Characterisation of the desiccant solution's thermophysical properties is essential for accurate modelling of the dehumidifier and regenerator, and for experimental evaluation of the desiccant system. The density and specific heat capacity are

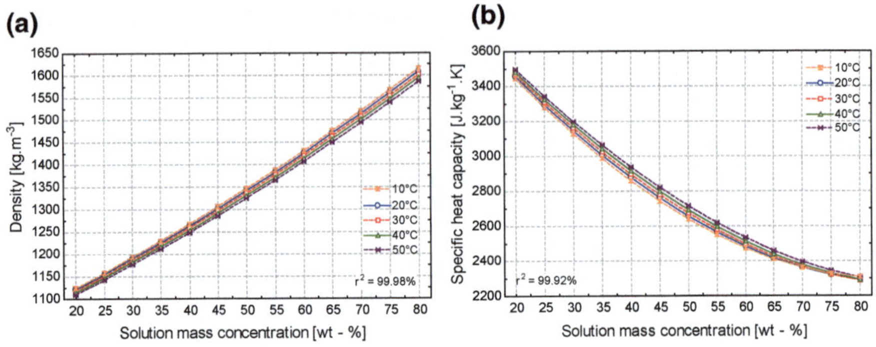

Fig. 3.5 **a** Density and **b** specific heat capacity plots of $CHKO_2$ solution

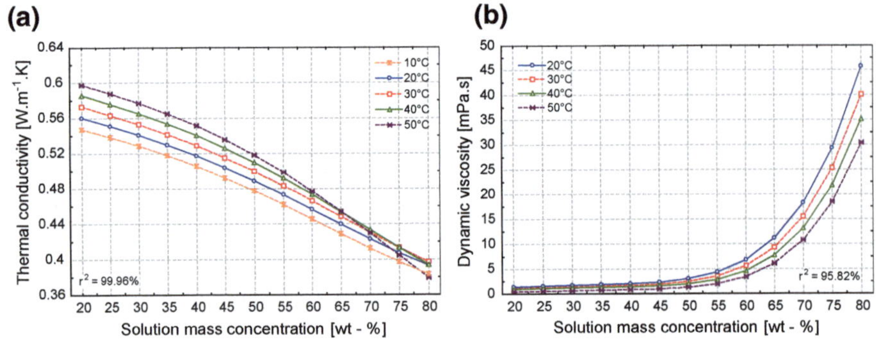

Fig. 3.6 **a** Thermal conductivity and **b** dynamic viscosity plots of $CHKO_2$ solution

shown in Fig. 3.5 and the thermal conductivity and dynamic viscosity are shown in Fig. 3.6. The data presented is for a temperature of 10–50 °C and solution mass concentration of 0.2–0.8 as this is the typical working range in a liquid desiccant air conditioning system. The $CHKO_2$ thermophysical property plots presented are linear regression curve fits to published empirical data from Melinder (2007). The coefficient of determination (r^2) is provided to demonstrate the fit between the statistical model and empirical data. The linear regression curve fit functions are provided in Appendix A (A1.2).

Section 3.2 has assessed three desiccant solutions, and has justified the selection of $CHKO_2$, based on a variety of criteria as the desiccant working fluid for the fuel cell tri-generation system. The operating conditions of the liquid desiccant solution in the dehumidifier have been selected as 20–30 °C at a solution mass concentration of 0.65–0.7, as this provides good performance over a range of climatic conditions. As highlighted in Chap. 2, there is limited published and available literature within the public domain regarding the specific physical characterisation of the $CHKO_2$ solution for liquid desiccant air conditioning applications. The work presented in Sect. 3.2 provides a contribution to knowledge with the assimilation of various sources of data to physically characterise the $CHKO_2$ solution in the required operating range for a liquid desiccant air conditioning application. The routines and curve fit functions can be referred to in Appendix A. Validation of the $CHKO_2$ dehumidifier/regenerator simulations with experimental results in Sect. 6.3 demonstrates the accuracy of the physical characterisation data presented. Next, Sect. 3.3 presents the equations and modelling procedure for accessing the dehumidifier and regenerator heat and mass exchangers.

3.3 Dehumidifier and Regenerator Mathematical Model

The aim of the modelling work is to investigate the performance of the dehumidifier and regenerator membrane heat and mass exchangers. This is to (1) enhance the knowledge of the performance of liquid desiccant air conditioning technology, (2) to offer specific considerations for the integration of liquid desiccant and SOFC technology in a tri-generation system set-up and (3) to facilitate the modelling of a SOFC tri-generation system in Chap. 4.

3.3.1 Mathematical Model Description

This section will describe the development of the mathematical model used to evaluate the dehumidifier and regenerator membrane heat and mass exchangers. The primary role of the dehumidifier and regenerator is to bring the desiccant solution into contact with an airstream without desiccant entrainment. The dehumidification and regeneration process involves the transfer of heat and mass, and are essentially the same but in reverse. Figure 3.7a shows a schematic diagram of the modelled heat and mass exchanger, whilst Fig. 3.7b shows a schematic diagram of a single air channel with the desiccant solution flowing either side of the air channel.

The modelled heat and mass exchanger shown in Fig. 3.7a consists of a series of channels that allow air and desiccant solution to flow in parallel, separated by a fibre membrane. The solution channels consist of polyethylene sheet, with fibre membranes attached on either side. The gap between the two solution channels provides the space for the air to flow. The modelling work assumes that during the dehumidification and regeneration process the desiccant solution completely wets the fibre membrane, and is in direct contact with the airstream. As a result, the membrane poses no resistances to heat and mass transfer, an assumption adopted by previous researchers (Jradi and Riffat 2014a). The model validation presented in Sects. 3.4 and 6.3 demonstrates good agreement with experimental data for a membrane based liquid desiccant heat and mass exchanger.

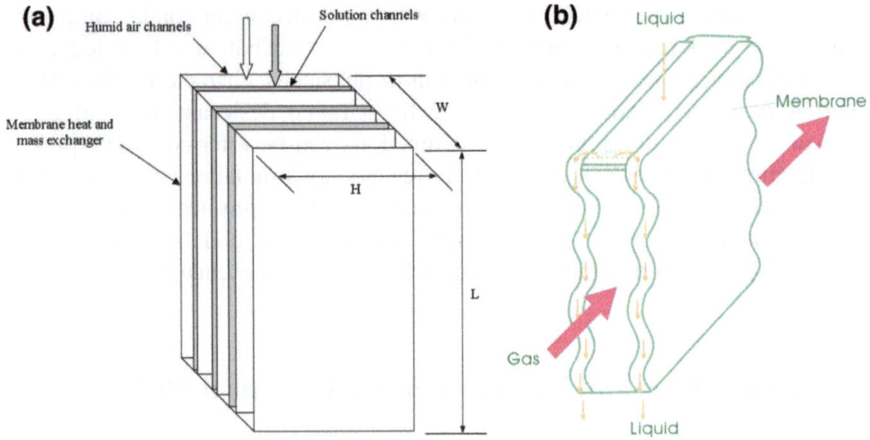

Fig. 3.7 **a** Schematic diagram of the membrane dehumidification heat and mass exchanger and **b** single channel heat and mass exchanger contacting design

In the dehumidifier, strong cool desiccant solution enters the top of the exchanger and flows downwards, wetting the membrane. The humid air stream passes across the wetted membrane in the adjacent air channel. At the membrane surface, differences in the vapour pressure of the desiccant solution and the humid air causes water vapour absorption from the humid airstream to the desiccant solution, resulting in a reduction in the absolute humidity of the airstream, and an increase or decrease in the air stream temperature depending on the desiccant solution temperature. The desiccant solution absorbs moisture, which reduces its concentration and increases its temperature due to the exothermic nature of the absorption process and heat transfer from the humid air stream.

Absorption of water vapour in the dehumidifier weakens the desiccant solution, therefore reducing its dehumidification potential. The regenerator is used to remove this added water vapour in the desiccant solution. Desorption is usually achieved by heating the desiccant solution to create a higher desiccant side vapour pressure and drive the moisture from the desiccant solution to a scavenger airstream, thus increasing the desiccant solution concentration. The scavenger air stream may be fresh or return room air depending on the application. The regeneration process increases the absolute humidity and temperature of the scavenger air stream.

The dehumidification and regeneration processes are very similar, relying on differences in vapour pressure between the desiccant solution and airstream to drive mass transfer. Therefore, the equations used to evaluate the heat and mass transfer processes in the dehumidifier are the same as those used for the regenerator, and are described together, the driving pressures are simply reversed.

In this modelling work, it is assumed that the heat and mass transfer processes are carried out in the laminar flow regime only. Table 3.3 shows the

Table 3.3 Reynolds number calculations for a single air channel

Air condition	T_a (°C)	RH_a (%)	Re_a
Hot and humid air	35	80	1107
Warm and dry air	25	50	1205

Reynolds (Re) number calculated for a single air channel, with dimensions of 0.21 m × 0.41 m × 0.01 m, at an air velocity of 1 m s^{-1}. Two air conditions are shown. The results show the air flow to be in the laminar regime i.e. <2300 for both cases, and thus the heat and mass transfer equations presented in Sects. 3.2.1 and 3.2.2 can be used with confidence.

The dehumidification process involves both heat and mass transfer; therefore the mathematical model developed needs to solve these sets of equations simultaneously.

3.3.1.1 Heat Transfer

Heat transfer is induced by the difference in temperature between the air stream and the surface of the desiccant soaked membrane, assumed to be equal to the bulk temperature of the desiccant solution.

For forced convection heat transfer in a closed conduit, the air-side heat transfer, *dq*, in J is determined using Eq. 3.1 (Welty et al. 2008).

$$dq_a = \frac{T_{a,in} - T_{sol,in}}{\frac{1}{k_a^h} + \frac{1}{k_{sol}^h}} SA_m \tag{3.1}$$

$T_{a,in}$ and $T_{sol,in}$ are the respective inlet temperatures of the air and desiccant solution in °C. k_a^h and k_{sol}^h are the respective convective heat transfer coefficients of the air and solution in W m^{-2} K. SA_m is the surface area available for heat and mass exchange between the air stream and desiccant solution in m^2.

The solution side heat transfer in J is defined in Eq. 3.2 (Liu 2008).The heat transferred to the solution includes the sensible heat due to the temperature difference between the air and desiccant plus the latent heat of water vapour absorption.

$$dq_{sol} = \frac{T_{a,in} - T_{sol,in}}{\frac{1}{k_a^h} + \frac{1}{k_{sol}^h}} SA_m + \left(2.501 \times 10^6 - 2370\ T_{sol,in}\right) dm \tag{3.2}$$

dm is the mass of water vapour condensed from the air to the desiccant solution in kg s^{-1}, and is defined in Eq. 3.8. The convective heat transfer coefficient in W m^{-2} K of the air and desiccant is defined in Eq. 3.3 (Çengel and Ghajar 2011).

$$k^h = 1.86\left(\mathrm{Re}\,\mathrm{Pr}\,\frac{D_h}{L}\right)^{1/3}\left(\frac{\mu_b}{\mu_w}\right)^{0.14}\frac{k^t}{D_h} \tag{3.3}$$

Re is the Reynolds number of the fluid. μ_b and μ_w are the dynamic viscosity in Pa s of the bulk air moving through the exchanger, and that of the desiccant solution at the exchanger wall respectively. k^t is the thermal conductivity of the fluid in W m^{-1} K. D_h is the hydraulic diameter of the channel in metres, and is defined in Eq. 3.6.

The Reynolds number is defined in Eq. 3.4 as:

$$\mathrm{Re} = \frac{\rho u D_h}{\mu} \tag{3.4}$$

ρ is the bulk density of the fluid in kg m^{-3}, u is the velocity of the fluid in m s^{-1}. The Prandtl (Pr) is defined in Eq. 3.5 as:

$$\mathrm{Pr} = \frac{c_p \mu}{k^t} \tag{3.5}$$

c_p is the specific heat capacity of the fluid in J kg^{-1} K
The hydraulic diameter, D_h, is defined in Eq. 3.6 as:

$$D_h = \frac{(2H_c w_c)}{H_c + w_c} \tag{3.6}$$

H_c and w_c are the height and width in metres of the channel, respectively.
The overall heat transfer coefficient (k_{all}^h) of the investigated heat and mass exchanger is calculated using Eq. 3.7 (Ge et al. 2014).

$$k_{all}^h = \left(\frac{1}{k_a^h} + \frac{1}{k_{sol}^h}\right)^{-1} \tag{3.7}$$

3.3.1.2 Mass Transfer

Mass transfer is driven by a vapour pressure differential between the air and desiccant solution. Equation 3.8 is used to determine the mass transferred in kg s^{-1} (Welty et al. 2008).

$$dm = \frac{p_v - p_{sol}}{R_w T_{a,in}\left(\frac{1}{k_a^m} + \frac{H}{k_{sol}^m}\right)} SA_m \tag{3.8}$$

where p_v and p_{sol} are the inlet equivalent vapour pressure of the air and liquid desiccant solution respectively in Pa. k_a^m and k_{sol}^m are the respective convective mass transfer coefficients of the air and desiccant solution in kg m^{-2} s. R_w is the molar gas constant for water (461.5 J kg^{-1} K). H is the Henry's Law constant—a measure of the solubility of a gas in a liquid (wt%), and is dependent on the solute, the solvent and the temperature, under atmospheric pressure (101,325 Pa). The H constant for the $CHKO_2$ solution $\left(H_{CHKO_2}\right)$ has been gained from experimental data presented in the literature (Stephen and Stephen 1963), the relationship to solution temperature is shown in Eq. 3.9.

$$H_{CHKO_2} = 104.3 - 0.3828\, T_{sol} \tag{3.9}$$

The convective mass transfer coefficient of the airstream in kg m^{-2} s can be determined from Eq. 3.10 (Çengel and Ghajar 2011).

$$k_a^m = k_a^h \rho_a c_{p,a} (Le)^{2/3} \tag{3.10}$$

$k_{\mathrm{a}}^{\mathrm{h}}$ is the convective heat transfer coefficient of the airstream in W m^{-2} K, determined using Eq. 3.3. Le is the Lewis number, and is defined in Eq. 3.11 as:

$$\mathrm{Le_a} = \frac{\mathrm{Sc_a}}{\mathrm{Pr_a}} \tag{3.11}$$

The Sc is the Schmidt number and is defined in Eq. 3.12 as:

$$\mathrm{Sc_a} = \frac{\mu_{\mathrm{a}}}{\rho_{\mathrm{a}} \mathrm{D_{AB}}} \tag{3.12}$$

The Prandtl number is previously defined in Eq. 3.5. D_{AB} is the mass diffusion coefficient, and is defined in Eq. 3.15.

The convective mass transfer coefficient in kg m^{-2} s of the desiccant solution is determined using Eq. 3.13 (Liu 2008).

$$h_{sol}^m = h_{sol}^h (\rho_{sol} c_{p,sol})^{-1} \left(\frac{\alpha_{sol}}{\mathrm{D_{AB}}} \right)^{-2/3} \tag{3.13}$$

$k_{\mathrm{sol}}^{\mathrm{h}}$ is the heat transfer coefficient of the airstream, determined using Eq. 3.3. The thermal diffusivity of the desiccant solution in m^2 s^{-1} is calculated using Eq. 3.14:

$$\alpha_{sol} = \frac{k_{sol}^t}{\rho_{sol} c_{p,sol}} \tag{3.14}$$

$k_{\mathrm{sol}}^{\mathrm{t}}$ is the thermal conductivity of the desiccant solution in W m^{-1} K. D_{AB} is the mass diffusion coefficient of water vapour into the desiccant solution and is a function or air temperature and pressure. The mass diffusion coefficient is expressed in Eq. 3.15 (Welty et al. 2008).

$$\mathrm{D_{AB}} = D_0 \frac{P_0}{P} \left(\frac{T_{sol,in}}{T_0} \right)^{3/2} \tag{3.15}$$

where $\frac{P_0}{P}$ is equal to one under atmospheric pressure conditions, D_0 is the binary diffusion coefficient of vapour in air = 2.46×10^{-5} m^2 s^{-1} and $T_0 = 273.15$ K.

The overall convective mass transfer coefficient in kg m^{-2} s of the investigated heat and mass exchanger is calculated using Eq. 3.16 (Ge et al. 2014).

$$k_{all}^m = \left(\frac{1}{k_a^m} + \frac{1}{k_{sol}^m} \right)^{-1} \tag{3.16}$$

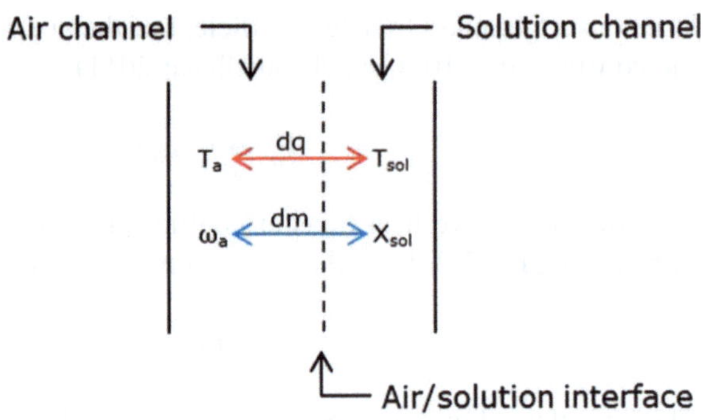

Fig. 3.8 The heat and mass transfer process in a single air/solution channel

Figure 3.8 provides a diagram conceptualising the flow of heat (*dq*) and mass (*dm*) in a single air and solution channel during the dehumidification and regeneration processes.

3.3.2 Computer Simulation

By applying heat and mass conservation, as shown in Eq. 3.17, it is possible to determine the outlet conditions of both the air and desiccant solution based upon the inlet boundary conditions, geometry of the heat and mass exchanger and calculated heat and mass transfer coefficients.

$$\dot{m}_a\left(h_{a,in} - h_{a,out}\right) = \dot{m}_{sol}\left(h_{sol,out} - h_{sol,in}\right) \tag{3.17}$$

$\dot{m}_{\mathrm{a}}$ and $\dot{m}_{\mathrm{sol}}$ are the respective air and solution mass flow rates in kg s^{-1}. The left-hand side of Eq. 3.17 represents the total enthalpy change (temperature and mass) of the air and the right-hand side represents the total enthalpy change of the desiccant solution. Engineering Equation Solver (EES) has been used to complete a one dimensional computer simulation to determine the thermal performance of the dehumidifier and regenerator. Some simplifying assumptions were made, and are as follows:

- The heat and mass transfer processes are considered steady-state.
- The dehumidification/regeneration processes are adiabatic; heat loss to surroundings is negligible.
- There is perfect wetting and coverage of the membrane material with desiccant solution, with ideal contact between the air and desiccant solution.
- The membrane material poses no resistance to heat and mass transfer i.e. direct contact is assumed.
- Air and desiccant solution physical and thermal properties are consistent between channels.

- No desiccant carry-over to the airstream occurred.
- The latent heat of sorption is absorbed by the desiccant solution.

The above assumptions have been adopted by previous researches in numerical studies, and have been found to have negligible effect on the overall desiccant system performance (Liu 2008; Jradi and Riffat 2014a).

The outlet air temperature is determined for each channel using Eq. 3.18.

$$T_{a,out} = T_{a,in} - \left(\frac{dq_{air}}{(1005 + 1820\omega_a)\dot{m}_a} \right) \tag{3.18}$$

dq_{air} is the heat transfer on the air side in J, and is defined in Eq. 3.1 and ω_{a} is air absolute humidity in $kg_{\text{vapour}}/kg_{\text{dryair}}$.
The outlet desiccant solution temperature is determined for each channel using Eq. 3.19.

$$T_{sol,out} = T_{sol,in} + \left(\frac{dq_{sol}}{\dot{m}_{sol} c_{p,sol}} \right) \tag{3.19}$$

dq_{solution} is the heat transfer on the solution side in J, and is defined in Eq. 3.2. The heat transferred to the solution includes the sensible heat due to the temperature difference plus the latent heat of water vapour absorption. dq_{air} and dq_{solution} will be positive during dehumidification and negative during regeneration.

The outlet air absolute humidity in $kg_{\text{vapour}}/kg_{\text{dryair}}$ is determined for each channel using Eq. 3.20

$$\omega_{a,out} = \frac{(\omega_{a,in} - dm)}{\dot{m}_a} \tag{3.20}$$

The outlet desiccant solution mass concentration is determined using Eq. 3.21

$$X_{sol,out} = X_{sol,in} \left(\frac{1}{1 + \left(\frac{\dot{m}_a}{\dot{m}_{sol}} \right) dm} \right) \tag{3.21}$$

dm is the mass of water vapour absorbed/desorbed in kg s^{-1} and is defined in Eq. 3.8. dm will be positive during dehumidification and negative during regeneration.
The liquid desiccant simulation work has been carried out using EES where air and water property routines are in-built validated functions. Appendix B provides the equations used to determine the thermodynamic properties of moist air as a reference. The thermophysical properties of the desiccant solution are determined from linear regression curve fits to published empirical data (James 1998; Melinder 2007). Section 3.3.3 provides the evaluation metrics used to analyse the performance of the dehumidifier and regenerator.

3.3.3 Performance Evaluation of Membrane Heat and Mass Exchanger

There are numerous evaluation methods presented in the literature for liquid desiccant based heat and mass exchangers. Below, the evaluation metrics used in the dehumidifier and regenerator parametric analysis are presented. The appraisal methods are also used in the experimental work presented in Chaps. 5–7. The metrics used are valid for both the dehumidification and regeneration processes.

The change in the absolute humidity (kg_{vapour}/kg_{dryair}) of air across the dehumidifier or regenerator is shown in Eq. 3.22.

$$\Delta\omega_{deh/reg} = \omega_{a,in} - \omega_{a,out} \tag{3.22}$$

A positive absolute humidity difference in the dehumidifier indicates water vapour has been removed from the air and transferred to the desiccant solution. A positive absolute humidity difference in the regenerator indicates water vapour has been removed from the desiccant solution and transferred to the air.

The rate of mass change (g s^{-1}) occurring in the dehumidifier is referred to as the moisture removal rate (MRR) and in the regenerator the moisture addition rate (MAR) and are shown in Eqs. 3.23 and 3.24 respectively.

$$\text{MRR} = \dot{m}_{a,deh}\left(\omega_{a,in,deh} - \omega_{a,out,deh}\right) \tag{3.23}$$

$$\text{MAR} = \dot{m}_{a,reg}\left(\omega_{a,out,reg} - \omega_{a,in,reg}\right) \tag{3.24}$$

$\dot{m}_a$ is the mass flow rate in kg s^{-1} of the air passing through the heat and mass exchanger. $\omega_{a,in}$ and $\omega_{a,out}$ are the respective inlet and outlet air absolute humidity in kg_{vapour}/kg_{dryair}.

The cooling output in W of the dehumidifier is shown in Eq. 3.25:

$$\dot{Q}_{cooling} = \dot{m}_a(h_{a,in} - h_{a,out}) \tag{3.25}$$

$h_{a,in}$ and $h_{a,out}$ are the respective inlet and outlet specific enthalpy of the air stream entering and leaving the heat and mass exchanger in J kg^{-1}. Air enthalpy is a function of both temperature and absolute humidity. Therefore air cooling means lowering temperature and/or absolute humidity i.e. reducing its enthalpy.

The latent heat transfer effectiveness, shown in Eq. 3.26, is the ratio of actual moisture transferred to the maximum moisture transfer.

$$\eta_{L,Deh/\text{Reg}} = \frac{\omega_{a,in} - \omega_{a,out}}{\omega_{a,in} - \omega_{eq}} \tag{3.26}$$

ω_{eq} is the equivalent moisture content in kg_{vapour}/kg_{dryair} of the desiccant solution at the inlet condition, and is a function of its concentration and temperature, shown in Eq. 3.27:

$$\omega_{eq} = 0.622\left(\frac{p_{sol}(X_{sol}, T_{sol})}{p_{atm} - p_{sol}(X_{sol}, T_{sol})}\right) \tag{3.27}$$

p_{sol} is the vapour pressure in Pa of the desiccant solution at a specified concentration and temperature. p_{atm} is atmospheric pressure and is equal to 101,325 Pa. X_{sol} is the desiccant solution mass concentration and T_{sol} is the solution temperature in °C.

The enthalpy effectiveness, shown in Eq. 3.28, is the ratio of actual air enthalpy change to the maximum change at ideal conditions.

$$\eta_{h,Deh/\mathrm{Reg}} = \frac{h_{a,in} - h_{a,out}}{h_{a,in} - h_{eq}} \tag{3.28}$$

h_{eq} is the air enthalpy in equilibrium with the liquid desiccant solution and is estimated using ω_{eq}, with an air temperature equal to that of the liquid desiccant solution.

Section 3.3 has introduced the mathematical model, computer simulation and performance evaluation metrics used for the dehumidifier and regenerator heat and mass exchangers. Next, Sect. 3.4 provides validation of the developed models.

3.4 Model Validation with Published Experimental Data

In order to use the developed model with confidence, the model needs to be validated with reliable experimental data. The model is first validated using published experimental data presented in the literature operating with a LiCl solution. The developed models operating with a $CHKO_2$ solution are then validated in Sect. 6.3. Ge et al. (2014) presents experimental data for a single liquid-to-air membrane based heat and mass exchanger, operating with a LiCl solution, for both the dehumidification and regeneration processes. A single heat and mass exchanger was used by the author so that a greater accuracy, control and measurement of the inlet and outlet conditions could be achieved. A high level of control of such factors makes the work well suited to model validation. The paper describes in detail the contactor design, experimental set up and testing method. The LiCl physical characteristics are gained from the property formulations presented by Conde (2009). Table 3.4 outlines the specification of the single heat and mass exchanger.

As previously stated in Sect. 3.1.1, in a desiccant air conditioning system, two fundamental processes occur: (1) dehumidification—moisture is absorbed by the

Table 3.4 Single heat and mass exchanger specification Ge et al. (2014)

Parameter	Value
Exchanger length (m)	0.99
Air channel height (m)	0.09448
Air to liquid contact area (m^2)	0.09405
Air gap thickness (m)	0.005
Solution gap thickness (m)	0.0012

desiccant solution, and (2) regeneration—moisture is desorbed from the desiccant solution. As a result, both the dehumidification and regeneration processes are validated with experimental data and potential reasons for discrepancies are listed in the validation summary in Sect. 3.4.3. The experimental data is shown alongside the simulation results. The authors do not state the error for the experimental results presented. Consequently, error bars of ±10 % have been included on the published results; this is to illustrate the degree of agreement between experimental and modelling data. As demonstrated in Chaps. 5 and 6, a ±10 % error margin is a reasonable assumption in the experimental testing of a liquid desiccant air conditioning system.

3.4.1 Dehumidification Validation

During the dehumidification process water vapour contained within the humid air stream is absorbed by the desiccant solution. Sensible cooling may also occur if the desiccant solution is at a temperature below that of the humid air stream. Table 3.5 shows the inlet conditions for twelve dehumidifier tests (D1–D12) carried out by Ge et al. (2014). These conditions were used as the inlet boundary conditions for the dehumidifier model during validation.

The performance indicators used for dehumidifier validation are outlet air absolute humidity, outlet air temperature, dehumidifier cooling, outlet desiccant solution mass concentration and latent and enthalpy effectiveness, as shown in Figs. 3.9, 3.10, and 3.11 respectively.

Over the 12 dehumidifier tests presented, the simulation results fit within ±10 % for all twelve cases based on outlet air temperature, seen in Fig. 3.9b, and desiccant solution mass concentration, seen in Fig. 3.10b. For outlet air absolute

Table 3.5 Inlet experimental conditions for LiCl dehumidifier tests Ge et al. (2014)

Test	m_a (kg s^{-1})	T_a (°C)	ω_a (kg$_{vap}$/kg$_{dryair}$)	m_{sol} (kg s^{-1})	T_{sol} (°C)	X_{sol} (–)
D1	0.001470833	30.1	0.0143	0.002308611	24	0.3208
D2	0.001018889	30	0.0143	0.0015675	24	0.32
D3	0.001502778	25	0.0142	0.002343056	24	0.3197
D4	0.001451111	35.2	0.0143	0.002264444	24	0.3204
D5	0.001475833	30	0.0202	0.002321111	24	0.3197
D6	0.001484722	30.1	0.0082	0.002277222	24	0.3197
D7	0.001478611	30	0.0145	0.001160833	24	0.3214
D8	0.001478056	30	0.0142	0.003460278	24	0.3214
D9	0.001480556	29.9	0.0138	0.002282222	28	0.3207
D10	0.001480556	30	0.0138	0.002279167	32	0.3201
D11	0.001478889	30	0.0142	0.002352222	24	0.3705
D12	0.001476667	30	0.0139	0.002188056	24	0.2765

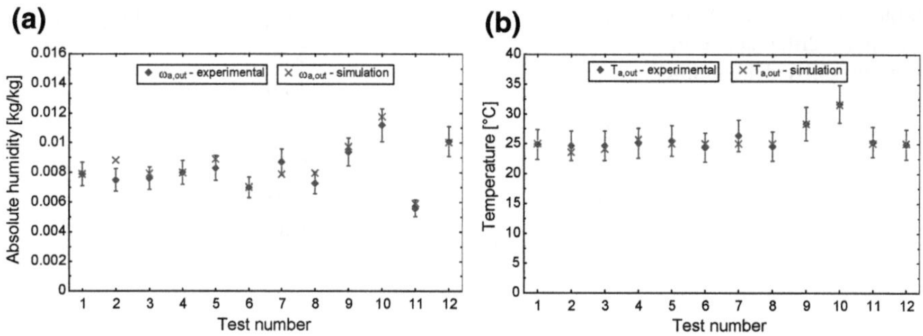

Fig. 3.9 **a** Dehumidifier outlet air absolute humidity and **b** dehumidifier outlet air temperature

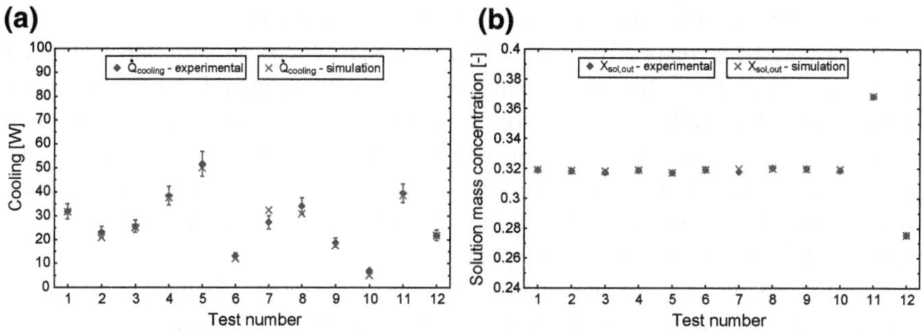

Fig. 3.10 **a** Dehumidifier cooling and **b** dehumidifier outlet solution mass concentration

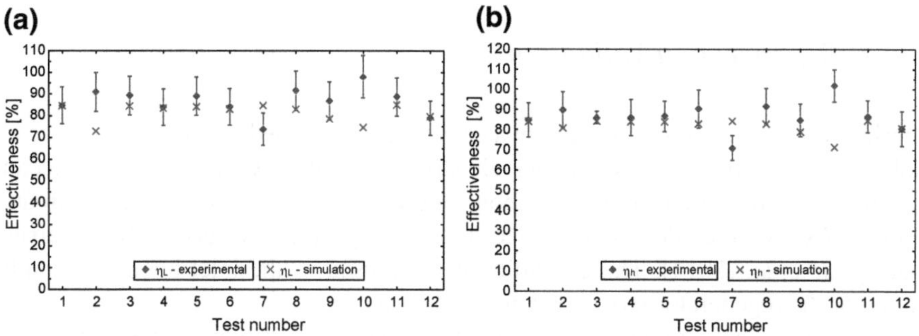

Fig. 3.11 **a** Dehumidifier latent effectiveness and **b** dehumidifier enthalpy effectiveness

humidity, Fig. 3.9a shows that all cases except test two, which has a 17 % difference, fits within the ±10 % margin. For cooling achieved, seen in Fig. 3.10a; tests seven and ten are outside of the ±10 % fit, and the difference is, 18 and 29 % respectively. For latent effectiveness, shown in Fig. 3.11a, tests two, seven and ten

Table 3.6 Average relative percentage difference results over the twelve dehumidifier tests

Parameter	Average relative difference (%)
$\omega_{a,out}$	5.36
$T_{a,out}$	1.89
$\dot{Q}_{cooling}$	7.78
$X_{sol,out}$	0.16
η_{L}	7.93
$\eta_{L,th}$	6.4
η_{h}	7.77

are outside of the ±10 % fit, the difference with experimental data is 19 14 and 23 % respectively. Ge et al. (2014) also present theoretical (*th*) values for the latent effectiveness for the twelve tests. When compared with the papers theoretical values, the difference with the simulation results for latent effectiveness is reduced to 19, 15 and 13 % respectively. It is apparent that where the simulation work is out by more than 10 %, the paper's theoretical results are also. For enthalpy effectiveness, Ge et al. (2014) do not present the theoretical values. The experimental enthalpy effectiveness values, presented in Fig. 3.11b, vary between 70 and 105 %, whilst the simulation values sit within the 70–85 % range. For enthalpy effectiveness, tests seven and ten are outside of the ±10 % fit, the difference with experimental data is 18 and 29 % respectively.

Table 3.6 shows the average relative percentage difference between the paper's experimental work and the simulations, for all performance parameters across the twelve tests. The average difference across all performance indicators is less than 10 %.

Next, Sect. 3.4.2 provides regenerator model validation.

3.4.2 Regeneration Validation

During the regeneration process water vapour contained within the desiccant solution is desorbed to a scavenger air stream, usually by heating the desiccant solution such that it has a higher vapour pressure than that of the inlet air stream. Table 3.7 shows the inlet conditions for thirteen regenerator tests (R1–R13) carried out by Ge et al. (2014). These conditions were used as the inlet boundary conditions for the regenerator model during validation.

The performance indicators used for the regenerator validation are outlet desiccant temperature, outlet desiccant solution mass concentration and latent and enthalpy effectiveness, as shown in Figs. 3.12 and 3.13 respectively. In terms of regenerator performance and its impact on the performance of the whole desiccant air conditioning system, the outlet conditions of the scavenger air stream are not critical. However, the outlet condition of the desiccant solution is, as this is what then flows back to the dehumidifier. As a result, the desiccant outlet condition is what has been focused on in the regenerator validation work.

Table 3.7 Inlet experimental conditions for LiCl regenerator tests Ge et al. (2014)

Test	m_a (kg s^{-1})	T_a (°C)	ω_a (kg$_{vap}$/kg$_{dryair}$)	m_{sol} (kg s^{-1})	T_{sol} (°C)	X_{sol} (–)
R1	0.001479444	30	0.0146	0.002237778	55.1	0.3216
R2	0.001017778	30	0.0137	0.001523889	55	0.3229
R3	0.000748611	30	0.0144	0.001113889	55	0.3229
R4	0.001505278	25	0.0142	0.0022775	55	0.3232
R5	0.001455278	35	0.0135	0.002198611	55	0.3232
R6	0.0014675	30	0.0199	0.002243889	55.1	0.3202
R7	0.001486667	30.1	0.0084	0.002206944	55	0.322
R8	0.001481667	29.9	0.0134	0.001123889	55	0.32
R9	0.001481389	30	0.0138	0.003326111	55	0.32
R10	0.001483611	29.9	0.0139	0.002241111	50.1	0.3204
R11	0.001483889	30	0.0142	0.002245833	45	0.3204
R12	0.001482778	30.1	0.0144	0.002265556	55.1	0.37
R13	0.00148	30	0.0143	0.0021375	55	0.27

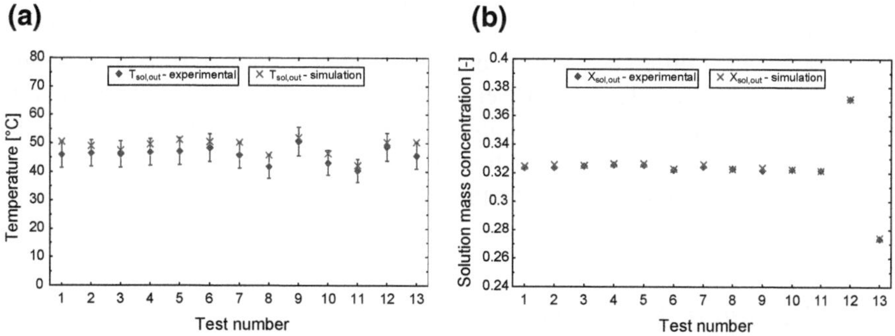

Fig. 3.12 **a** Regenerator outlet desiccant temperature and **b** regenerator outlet solution mass concentration

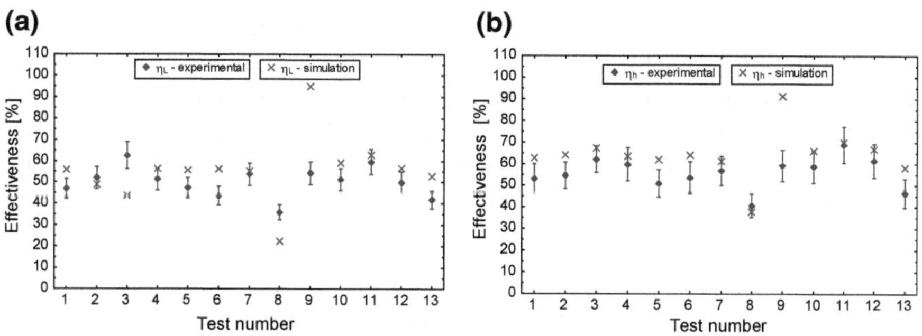

Fig. 3.13 **a** Regenerator latent effectiveness and **b** regenerator enthalpy effectiveness

Table 3.8 Average percentage difference results over the thirteen regenerator tests

Parameter	Average relative difference (%)
$T_{sol,out}$	6.6
$X_{sol,out}$	0.28
η_L	22.14
$\eta_{L,th}$	18.08
η_h	16.22

Figure 3.12a, b show that the simulated solution outlet temperature and desiccant solution mass concentration respectively, fit within the 10 % margin for all thirteen experimental tests. However, Fig. 3.13a, b show that the latent and enthalpy effectiveness do not fit the experimental data so well. Ge et al. (2014) explains that during regeneration, crystallisation of the desiccant solution at certain concentrations and temperature could pose a barrier to heat and mass transfer across the membrane, evident in the lower experimental latent and enthalpy effectiveness values compared to the simulation results. The average difference with the theoretical latent effectiveness values is lower.

Table 3.8 shows the average relative percentage difference between the paper's experimental work and the simulations, for all regenerator performance parameters. The average difference for outlet solution temperature and mass concentration is acceptable i.e. <10 %; however for the latent and enthalpy effectiveness the values show larger discrepancy. As highlighted by Ge et al. (2014), this difference can be attributed to the issue of solution crystallisation.

Next, Sect. 3.4.3 provides a validation summary.

3.4.3 Validation Summary

Sections 3.4.1 and 3.4.2 have validated to a suitable degree the developed dehumidifier and regenerator models. Experimental data for a membrane based dehumidifier/regenerator HMX operating with a LiCl desiccant solution has been published by Ge et al. (2014). The experimental results have been compared with the developed model. Overall the model seems to be working within an acceptable error range for the variables required to assess a desiccant air conditioning system. The model can therefore be used with confidence in the parametric analysis presented in Sect. 3.5 and the tri-generation system model in Chap. 4.

Potential reasons for the discrepancies between experimental and simulations have been identified for both the dehumidifier and regenerator, these are:

- Average values of experimental conditions were used as the modelling input.
- The air flow across the membrane creates a small angle with the falling desiccant solution in the experimental testing; however, for simplicity it was assumed parallel in the modelling work.

- Total coverage and full wetting of the membrane material with the desiccant solution may not have been achieved in the experiments, whereas it was assumed in the modelling.
- Perfect contact between the air and desiccant solution may not have been achieved in the experiments, whereas it was assumed in the modelling.
- The authors do not state the experimental error for the performance variables used in model validation, therefore the actual difference between experimental and modelling may be less, and within the $\pm 10\ \%$ margin.
- Measurement of the exact desiccant solution mass concentration is a challenging aspect of the experimental work and as demonstrated in Sect. 3.5, a small variation in the desiccant solution mass concentration has a large influence on dehumidifier/regenerator performance.

Section 3.4 has experimentally validated the desiccant dehumidifier and regenerator models to within an acceptable range. Experimental validation with a $CHKO_2$ solution is presented in Sect. 6.3. Next, using the validated models, Sect. 3.5 provides a parametric analysis for both the dehumidifier and regenerator.

3.5 Dehumidifier and Regenerator Parametric Analysis

Section 3.4 has validated, to an acceptable level, the dehumidifier and regenerator models, operating with a LiCl solution. This section presents a parametric analysis for the desiccant dehumidifier and regenerator, operating with a $CHKO_2$ solution.

The modelling work in this thesis does not aim to progress the state of the art, but to facilitate a better understanding of the process and performance of such systems and thus draw relevant conclusions that are pertinent to the integration of liquid desiccant air conditioning technology with a SOFC in a novel tri-generation system set-up. The parametric analysis will investigate both the dehumidification and regeneration processes, focusing on the impact inlet environmental and operating conditions have on performance.

The geometric specification and operating values for the dehumidifier and regenerator parametric analysis are presented in Table 3.9. The dimensions of the modelled heat and mass exchanger are based upon the experimental dehumidifier and regenerator (SDCS) presented in Chap. 6. The desiccant solution mass flow rate selected is based on a value that provides acceptable wetting of the membrane in the experimental system, without creating superfluous desiccant flow into the supply airstream. Often, theoretical investigations do not consider the practical implications of the operating values used. In liquid desiccant air conditioning systems high liquid desiccant flow rates result in unwanted desiccant solution carry-over to the supply airstream. Furthermore, high volumetric air flow rates are not achievable due to pressure drops in the exchanger core.

In the parametric investigations, a high humidity scenario has been selected to demonstrate the capability of liquid desiccant dehumidification systems in reducing air moisture content. Next, Sect. 3.5.1 presents the dehumidifier parametric analysis.

Table 3.9 Dehumidifier and regenerator geometrical specification and operating values

Parameter	Value
Length of exchanger (mm)	410
Height of exchanger (mm)	210
Width of air channel (mm)	10
Wall thickness (mm)	1
Inlet air temperature (°C)	30
Inlet air relative humidity (%)	85
Inlet air absolute humidity (kg_{vapour}/kg_{dryair})	0.02297
Desiccant solution volumetric flow (L min^{-1})	3.2
Desiccant solution mass flow rate per channel (kg s^{-1})	0.00192

Table 3.10 Dehumidifier base simulation conditions

Parameter	Value
Inlet air velocity (m s^{-1})	1.2
Inlet air volumetric flow (m^3 h^{-1})	191
Desiccant solution mass concentration (–)	0.7
Desiccant solution temperature (°C)	24
Convective heat transfer coefficient (W m^{-2} K)	11.4
Convective mass transfer coefficient (kg m^{-2} s)	0.0082

3.5.1 Dehumidifier Parametric Analysis

The dehumidifier parametric analysis will investigate the effect: inlet air relative humidity, inlet air temperature, inlet air velocity, desiccant solution mass flow rate and desiccant solution mass concentration have on the overall performance of the dehumidifier. These variables have been selected due to their influence on overall dehumidifier performance. The dehumidifier parametric analysis is based upon the information listed in Tables 3.9 and 3.10. The volumetric air flow rate has been selected based on achieving a change in absolute humidity across the dehumidifier of 0.01 kg/kg; an acceptable value for air conditioning applications. The convective heat and mass transfer coefficients listed are for the stated boundary conditions.

Figure 3.14 demonstrates the dehumidification process on a psychrometric chart. Based on the dehumidifier base simulation conditions, the process air is conditioned to a temperature of 26 °C and 61 % RH. The reduction in absolute humidity of the process air across the dehumidifier is 0.01006 kg_{vapour}/kg_{dryair}, an acceptable value for air conditioning applications (Jradi and Riffat 2014a). At the dehumidifier outlet the desiccant solution temperature is 25.3 °C, with a solution mass concentration of 0.6948. The overall dehumidification and cooling capacity is 1772 W, at a latent and enthalpy effectiveness of 44.93 and 47.1 % respectively.

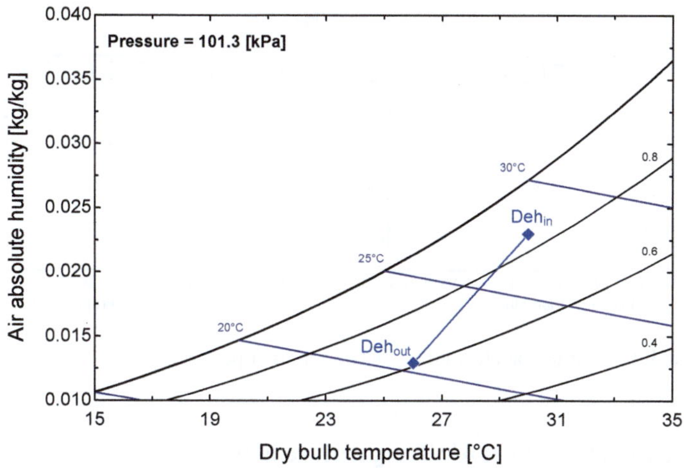

Fig. 3.14 The dehumidification process shown on a psychrometric chart

3.5.1.1 Dehumidifier Inlet Air Relative Humidity Effect

Figure 3.15 presents the effect varying the inlet air relative humidity from 40 to 90 % has on the performance of the dehumidifier. Figure 3.15a, demonstrates that the increase in air absolute humidity difference and cooling output is proportional to the inlet air relative humidity. As the inlet air relative humidity increases, so does the partial vapour pressure of the air, leading to a greater vapour pressure differential between the air and desiccant solution. An increase in the vapour pressure differential leads to increased mass transfer and thus a greater change in the absolute humidity of supply air, creating a larger (latent) cooling effect. Figure 3.15b demonstrates that increasing the inlet air relative humidity from 40 to 90 % increases both the latent and enthalpy effectiveness from 28.14 and 35.56 to 45.74 and 47.72 % respectively. Figure 3.15b demonstrates that the desiccant solution outlet mass concentration is inversely proportional to inlet air relative humidity.

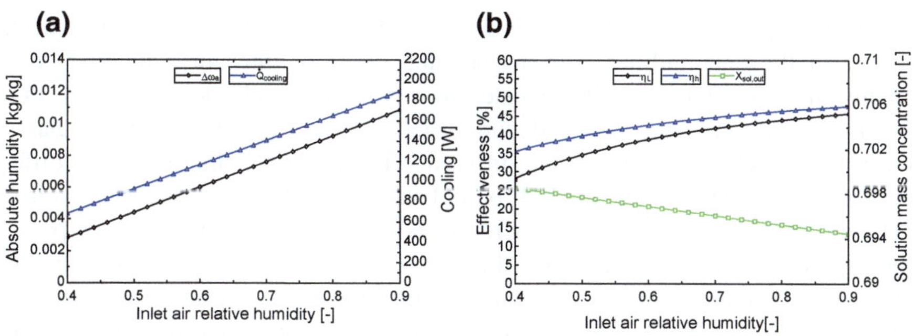

Fig. 3.15 **a, b** Inlet air relative humidity effect on dehumidifier performance

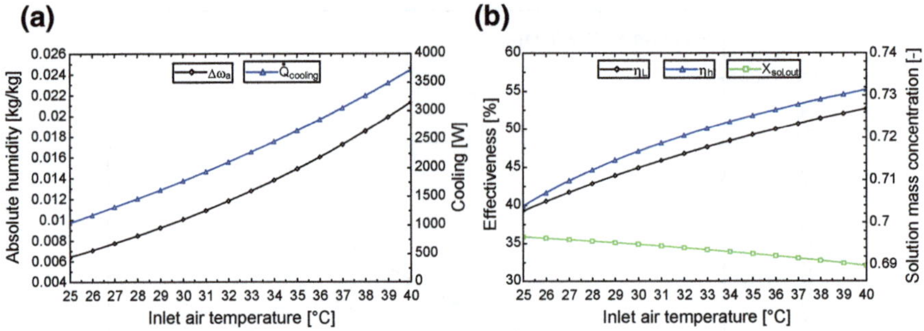

Fig. 3.16 **a, b** Inlet air temperature effect on dehumidifier performance

3.5.1.2 Dehumidifier Inlet Air Temperature Effect

Figure 3.16 presents the effect varying the inlet air temperature from 25 to 40 °C has on the performance of the dehumidifier. Figure 3.16a demonstrates that increasing supply air temperature produces a greater air absolute humidity difference and cooling output. At a constant relative humidity, an increase in supply air temperature results in a higher inlet air vapour pressure, resulting in an increased mass transfer potential. Figure 3.16b demonstrates that increasing the inlet air temperature from 25 to 40 °C increases both the latent and enthalpy effectiveness from 39.31 and 39.94 to 52.7 and 55.2 % respectively. Figure 3.16b demonstrates that the desiccant solution outlet mass concentration is almost inversely proportional to the increase in inlet air temperature. This is because as the as the inlet air temperature increases the potential for mass transfer increases.

It is evident that over the investigated range, the desiccant dehumidifier performance is more influenced by inlet air relative humidity changes than temperature, illustrated in the greater change in latent effectiveness.

3.5.1.3 Dehumidifier Inlet Air Velocity Effect

Figure 3.17 presents the effect varying the inlet air velocity has on the performance of the dehumidifier. The investigated range of inlet air velocity in each channel is 0.7–2.5 m s^{-1}, this equates to a volumetric air flow range of 111–397 m^3 h^{-1} through the whole dehumidifier. Figure 3.17a demonstrates that as the air inlet velocity increases the difference in the air absolute humidity difference reduces, however, the cooling output increases. With constant dehumidifier geometry, as the inlet air velocity increases it means a greater mass of air is passed through the dehumidifier, and thus the absolute humidity of supply air is higher. However, the cooling output increases because there is a greater mass of conditioned air. Figure 3.17b demonstrates that increasing the inlet air velocity from 0.7 to 2.5 m s^{-1} reduces both the latent and enthalpy effectiveness from 77.03 and

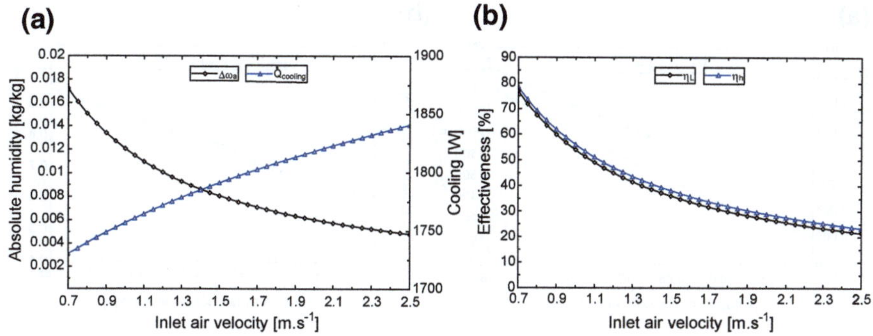

Fig. 3.17 **a**, **b** Inlet air velocity effect on the dehumidifier performance

78.82 to 21.57 and 23.49 % respectively. The desiccant solution outlet mass concentration remains constant.

3.5.1.4 Dehumidifier Desiccant Solution Mass Flow Rate Effect

Figure 3.18 presents the effect varying the inlet desiccant solution mass flow rate in each channel has on the performance of the dehumidifier. The investigated range of desiccant solution mass flow rates per channel is 0.0001–0.003 kg s^{-1}, this equates to a volumetric desiccant flow range of 0.16–5 L min^{-1} through the whole dehumidifier. Figure 3.18a demonstrates that as desiccant solution mass flow rate increases, the air absolute humidity difference increases, and the cooling output also increases. Figure 3.18b demonstrates that increasing the desiccant solution mass flow rate in each channel from 0.0001 to 0.003 kg s^{-1} increases both the latent and enthalpy effectiveness from 16.78 and 21.56 to 52.14 and 53.61 % respectively. Figure 3.18b demonstrates an increase in desiccant solution outlet mass concentration with increasing desiccant solution mass flow rate, with a respective sharp increase/decrease at very low flow rates (less than 0.8 L min^{-1}).

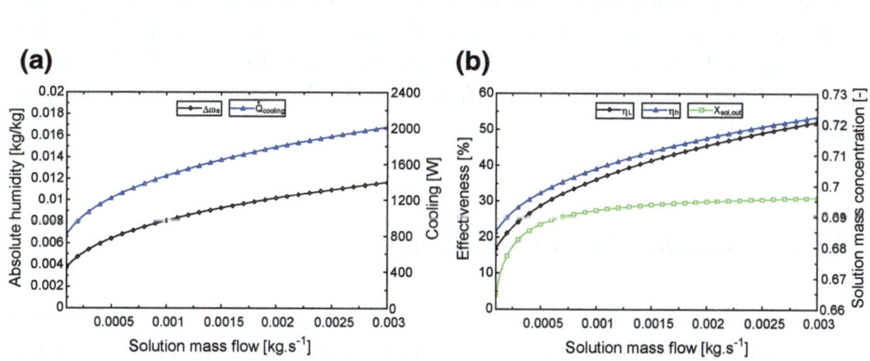

Fig. 3.18 **a**, **b** Desiccant solution mass flow rate effect on dehumidifier performance

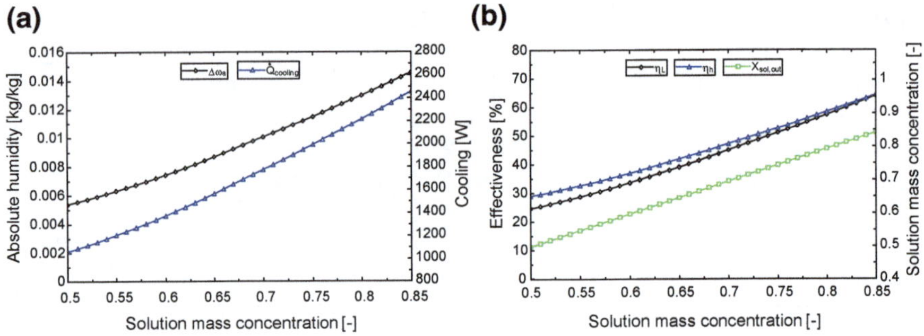

Fig. 3.19 **a, b** Desiccant solution mass concentration effect on dehumidifier performance

3.5.1.5 Dehumidifier Desiccant Solution Mass Concentration Effect

Figure 3.19 presents the effect varying the inlet desiccant solution mass concentration from 0.5 to 0.85 has on the performance of the dehumidifier. Figure 3.19a demonstrates a greater air absolute humidity difference and cooling output with increasing desiccant solution mass concentration. At a constant solution temperature an increase in desiccant solution mass concentration results in a lower solution vapour pressure, as shown in Fig. 3.3. Therefore, at a constant inlet air condition the vapour pressure difference between the inlet air and desiccant increases, and thus the mass transfer potential is enhanced. Figure 3.19b demonstrates that increasing the desiccant solution mass concentration from 0.5 to 0.85 increases both the latent and enthalpy effectiveness from 24.63 and 28.85 to 64.1 and 64.39 % respectively.

Section 3.5.1 has presented a parametric analysis investigating the effect: inlet air relative humidity, inlet air temperature, inlet air velocity, desiccant solution mass flow rate and desiccant solution mass concentration have on the performance of the dehumidifier membrane heat and mass exchanger. Based on the set boundary conditions, the desiccant dehumidifier can supply air at a suitable condition for air conditioning applications. The inclusion of a subsequent indirect evaporative cooling device could be considered for further sensible cooling of the air. Working at a wet bulb effectiveness of 78 % (Jradi and Riffat 2014b), a supply air temperature of 19.74 °C could be achieved. Next, Sect. 3.5.2 presents a regenerator parametric analysis.

3.5.2 Regenerator Parametric Analysis

The regenerator parametric analysis has a different perspective to that presented in the dehumidifier analysis. For the dehumidifier, the outlet condition of both the

Table 3.11 Regenerator base simulation conditions

Parameter	Value
Inlet air velocity (m s^{-1})	1.2
Inlet air volumetric flow (m^3 h^{-1})	191
Desiccant solution mass concentration (–)	0.6948
Desiccant solution temperature (°C)	58.5
Convective heat transfer coefficient (W m^{-2} K)	11.26
Convective mass transfer coefficient (kg m^{-2} s)	0.0072

air and desiccant solution are of importance. However, in the regenerator the air outlet condition is not critical but the outlet condition of the desiccant solution is, as this is what then flows back to the dehumidifier and governs whether the system is balanced. As a result, the desiccant outlet condition is what has been focused on in the regenerator parametric analysis. The regenerator parametric analysis will investigate the impact: inlet air relative humidity, inlet air temperature, desiccant solution mass flow, desiccant solution temperature and desiccant solution mass concentration have on the overall performance of the regenerator. These variables have been selected due to their influence on regenerator performance.

The regenerator parametric analysis uses the same geometrical specification and operating values used in dehumidifier analysis; these are listed in Table 3.9. Specific regenerator boundary conditions are provided in Table 3.11.

To facilitate continuous operation of a desiccant air conditioning system, the weak desiccant solution leaving the dehumidifier must be regenerated i.e. remove the moisture absorbed in the dehumidifier, and thus restore its dehumidification potential. The regenerator has been simulated in order to allow the continuous operation of the dehumidifier presented in Sect. 3.5.1. In Sect. 3.5.1, it was shown that the desiccant solution leaving the dehumidifier has a mass concentration of 0.6948 and temperature of 25.3 °C. The regenerator model shows that the desiccant solution must be heated to 58.5 °C to regenerate the desiccant solution back to a mass concentration of 0.7. Thus, in the regenerator parametric analysis, the inlet desiccant solution mass concentration and temperature have been set to 0.6948 and 58.5 °C respectively.

Figure 3.20 demonstrates the regeneration process on a psychrometric chart. Based on the regenerator conditions selected, the outlet regenerator air stream is heated to 48.61 °C with a 44.2 % RH. This equates to an increase in the air streams absolute humidity of 0.00994 kg_{vapour}/kg_{dryair}. At the regenerator outlet the desiccant solution temperature is 52.43 °C, with a mass concentration of 0.7. The latent effectiveness of this process is 34.44 %.

Following regeneration the desiccant solution needs to be sensibly cooled prior to entering the dehumidifier; this could be achieved in two stages (1) in a desiccant to desiccant plate heat exchanger, exchanging heat with the desiccant flowing to the regenerator, and (2) using cooling water from an evaporative cooling process.

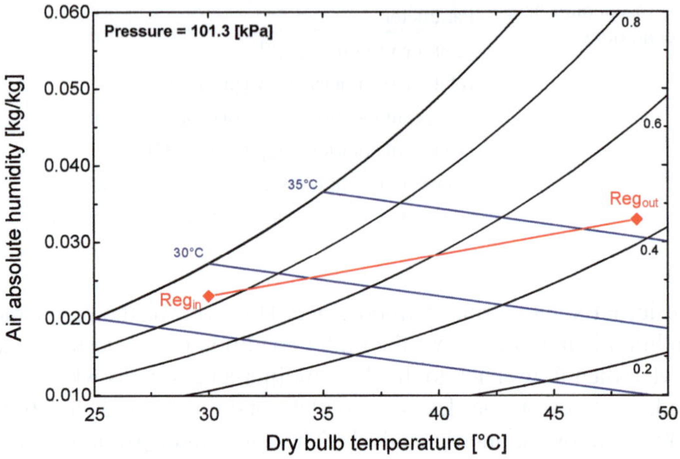

Fig. 3.20 The regeneration process shown on a psychrometric chart

3.5.2.1 Regenerator Inlet Air Relative Humidity Effect

Figure 3.21 presents the effect varying the inlet air relative humidity from 20 to 80 % has on the performance of the regenerator. Figure 3.21a demonstrates that the outlet solution temperature and concentration are inversely proportional to inlet air relative humidity. Furthermore, Fig. 3.21b demonstrates that as the relative humidity of inlet air increases, the air absolute difference and latent effectiveness reduces. This is due to the fact that as relative humidity increases, the vapour pressure of the air increases. This results in a smaller vapour pressure differential between the solution and air, and thus less mass transfer. Lower inlet air relative humidity means the desiccant solution at a specified inlet condition can be regenerated to a higher concentration.

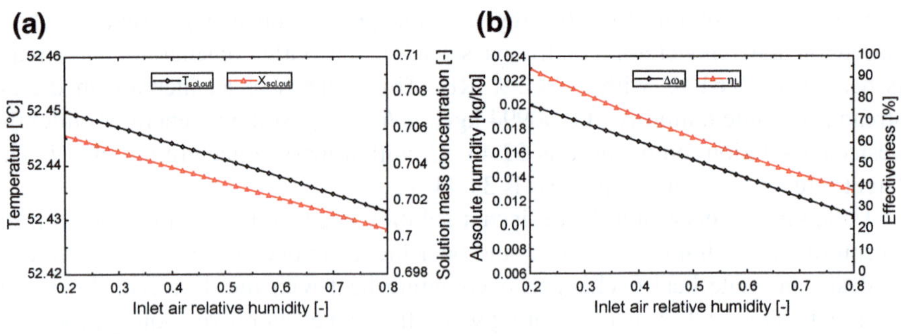

Fig. 3.21 **a**, **b** Inlet air relative humidity effect on regenerator performance

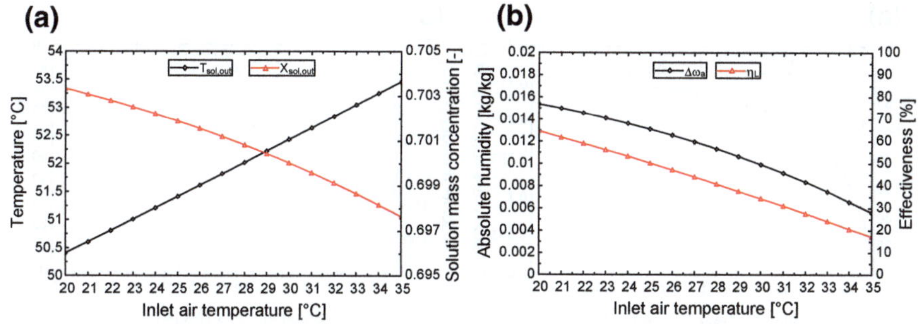

Fig. 3.22 **a, b** Inlet air temperature effect on regenerator performance

3.5.2.2 Regenerator Inlet Air Temperature Effect

Figure 3.22 presents the effect varying the inlet air temperature from 20 to 35 °C has on the performance of the regenerator. Similar to the relative humidity effect, the vapour pressure of the inlet air increases with rising temperature. Figure 3.22a demonstrates that the outlet desiccant solution mass concentration is lower with increasing inlet air temperature i.e. it is re-concentrated to a lesser degree. Figure 3.22b demonstrates that the air absolute humidity difference and latent effectiveness reduces as inlet air temperature increases.

Although good cooling capacity has been demonstrated in hot and humid climatic conditions, regeneration capacity reduces with increasing inlet air relative humidity and temperature. The successful application of liquid desiccant air conditioning is therefore location and climatic specific. When operating the regenerator on fresh outside air, liquid desiccant system performance will be poorer in hot and high humid climates.

In this parametric study, the inlet air condition to the regenerator has been assumed to be equal to the fresh air used in the dehumidifier. However, as highlighted in Sects. 3.5.2.1 and 3.5.2.2 regeneration capacity is higher with a lower inlet air relative and temperature. In a building application, fresh conditioned air is supplied to the space from the dehumidifier. The mass of supply air must be balanced with extract to remove internal gains and prevent pressurisation of the space. The extracted room air will be stale, thus it cannot be used again for human comfort purposes. However it can be used for regeneration purposes. This is favourable as it will have a lower temperature and relative humidity than the outside fresh air, thus improving the performance of the regeneration process (Dwyer 2014).

3.5.2.3 Regenerator Desiccant Solution Mass Flow Rate Effect

Figure 3.23 presents the effect varying the inlet desiccant solution mass flow rate in each channel has on the performance of the regenerator. The investigated

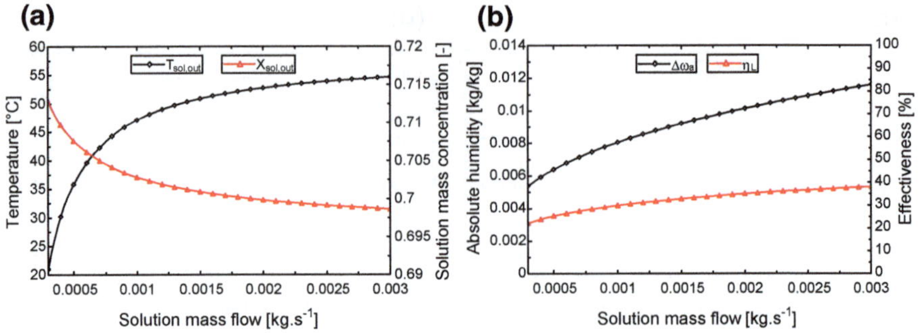

Fig. 3.23 **a, b** Desiccant solution mass flow rate effect on regenerator performance

range of desiccant solution mass flow rates per channel is 0.0003–0.003 kg s^{-1}, this equates to a volumetric desiccant flow range of 0.5–5 L min^{-1} through the whole regenerator. Figure 3.23a demonstrates that as the solution mass flow rate increases above 0.001 kg s^{-1}, the outlet solution temperature increases, and the outlet solution concentration decreases. Figure 3.23b demonstrates that the air absolute humidity difference and latent effectiveness both increase with increasing solution mass flow rate. In order for the desiccant air conditioning system to be balanced, the mass condensed in the dehumidifier needs to be evaporated off in the regenerator. Therefore a desiccant solution mass flow rate must be selected such that this is satisfied. In the theoretical analysis, the desiccant solution flow rate is equal in both the dehumidifier and regenerator. Mass balance is achieved by heating the desiccant solution to an adequate temperature. This is discussed next in Sect. 3.5.2.4.

3.5.2.4 Regenerator Desiccant Solution Temperature Effect

Figure 3.24 shows the impact increasing desiccant solution inlet temperature has on the performance of the regeneration process. Figure 3.24a demonstrates that as the desiccant solution temperature is increased, the outlet concentration rises. This is because a higher solution temperature creates a higher vapour pressure, and thus a greater vapour pressure differential between the regenerator airstream, resulting in a greater mass of vapour transferred from the desiccant solution to the air. As the inlet desiccant solution temperature increases, so does the outlet solution temperature. Figure 3.24b demonstrates, for the stated inlet conditions, the air absolute humidity difference and latent effectiveness increases with solution temperature. It is evident that the regeneration process, indicated by a positive absolute humidity difference/latent effectiveness, begins at an inlet solution temperature of 47.5 °C at a 0.6948 solution mass concentration. This is a realistically achievable temperature when integrated with an SOFC CHP system. Above 47.5 °C the regeneration capacity increases. Re-concentration of the desiccant

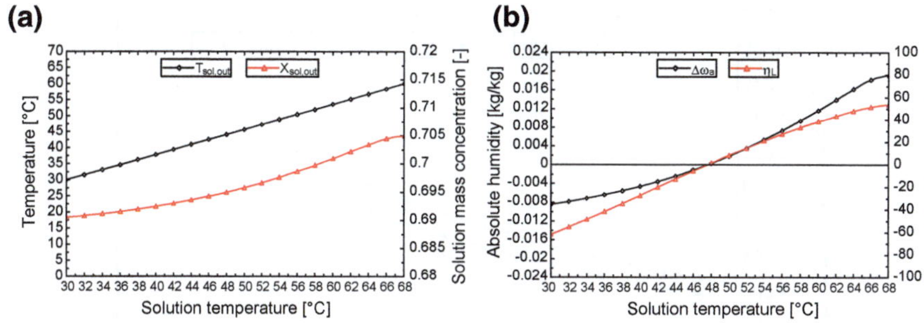

Fig. 3.24 a, b Desiccant solution temperature effect on regenerator performance

solution back to the dehumidifier mass concentration of 0.7 is achieved at an inlet solution temperature of 58.5 °C. Above 58.5 °C, the desiccant solution is re-concentrated above 0.7.

A higher inlet solution temperature means greater re-concentration of the desiccant solution, representing a greater cooling potential in the dehumidifier. However a higher inlet solution temperature means a greater thermal energy input is required, potentially lowering the COP_{th} of the system. Furthermore, in a desiccant air conditioning system the outlet solution from the regenerator needs to be cooled to an acceptable level prior to it re-entering the dehumidifier. If the solution outlet temperature from the regenerator is too high this may (a) incur a large energy penalty for cooling, or (b) not be possible with the available means i.e. evaporatively. It is therefore critical not to operate the regenerator at too high a temperature as this can have an adverse impact on the total system performance.

Figure 3.25 presents a plot of the change in absolute humidity ($\Delta\omega_a$) of the regenerator airstream with respect to inlet desiccant solution mass concentration and temperature. It is evident that the regeneration process, indicated by a positive absolute humidity difference (air is taking on water vapour from the desiccant solution) begins at a lower inlet solution temperature at lower solution concentrations. This is a very important observation when considering the integration of a liquid desiccant air conditioning system in a tri-generation system set-up. The expected flow temperature of the water in the WHR loop of the prime mover, i.e. SOFC, will dictate what solution concentration the desiccant system should be operated at. If suitable matching is not achieved, the liquid desiccant system may not operate in a balanced manner i.e. mass balance in the dehumidifier and regenerator. Regeneration of the $CHKO_2$ solution begins at 44.5 °C for a 0.65 solution mass concentration, 48 °C for a 0.7 solution mass concentration and 52 °C for a 0.75 solution mass concentration.

Section 3.5.2 has presented a regenerator parametric analysis investigating the effect inlet air relative humidity, inlet air temperature, desiccant solution mass flow rate, desiccant solution temperature and desiccant solution mass concentration have on the performance of the regenerator. Using the input conditions described in the dehumidifier and regenerator parametric studies, along with a desiccant to

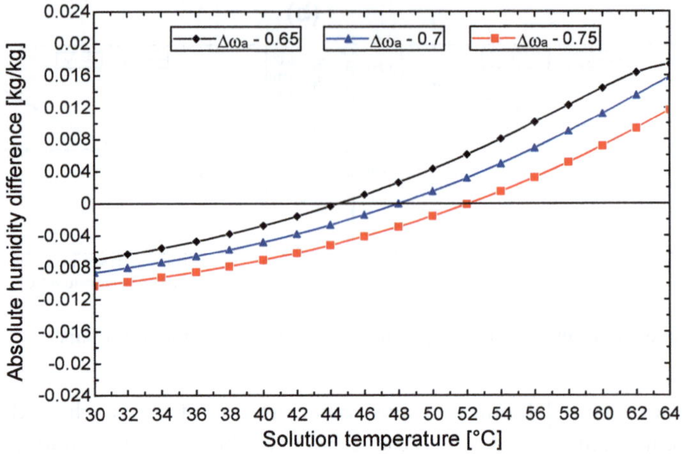

Fig. 3.25 Desiccant solution mass concentration and temperature effect on regenerator performance

desiccant heat exchanger, the desiccant air conditioning system achieves a COP_{th} and COP_{el} of 0.65 and 16.1 respectively. The liquid desiccant air conditioning system electrical requirement of 110 W has been gained from an experimental investigation presented in Chap. 6. The COP_{th} and COP_{el} calculation method is provided in Eqs. 4.30 and 4.31 respectively. Conclusions from both the dehumidifier and regenerator parametric analysis are summarised in Sect. 3.5.3, with particular reference to their importance to the SOFC tri-generation system.

3.5.3 Parametric Analysis Conclusions and Implications for Tri-generation System Integration

This section aims to summarise the conclusions from the dehumidifier and regenerator parametric analyses presented in Sects. 3.5.1 and 3.5.2 respectively. In order to fulfil the chapter aims set out in Sect. 3.1, particular reference to the importance to liquid desiccant and tri-generation system integration and performance is made throughout.

The investigated dehumidifier operates in a latent and enthalpy effectiveness range of 33–53 and 38–56 % respectively. Values typical of a membrane based heat and mass exchanger reported in the literature. Dehumidification capacity is greater in hotter more humid conditions, making the technology well suited to humid subtropical/continental climates such as South East Asia and the Mediterranean. Latent and enthalpy effectiveness are directly proportional to the intake air temperature, intake air relative humidity and liquid desiccant flow rate, but inversely proportional to the intake air velocity. As dehumidification potential increases, the dilution of the liquid desiccant solution becomes greater. In order to restore the desiccant solution back to its original working concentration, and to keep the system balanced, the regeneration requirement will be larger. In a tri-generation system set-up, the regenerator capacity and thus dehumidifier capacity, will be a function of the available thermal output from the prime mover, in this case the SOFC. Evaporative based cooling is an appropriate technological pairing for both desiccant solution and supply air cooling. This is due to its requirement to operate on hot dry air and its low electrical demand.

Regeneration capacity increases as the temperature and relative humidity of the inlet air decreases. The successful application of liquid desiccant air conditioning is very much location and climatic specific. When operating the regenerator on fresh outside air, liquid desiccant system performance will be poorer in hot and high humid climates. It is therefore of benefit to system performance to operate the regenerator on the return room air from the building. In this scenario, compared to using fresh outside air, a lower thermal input can be used to produce the same regeneration capacity. It is evident that increasing the desiccant solution mass flow rate in the dehumidifier generates a greater cooling potential. However, a greater desiccant solution mass flow in the regenerator means greater thermal input is needed to heat the solution to the required temperature. Therefore the desiccant solution mass flow rate in the SOFC tri-generation system will be influenced to some degree by the available thermal output from the SOFC. As the operating concentration of $CHKO_2$ solution increases, the required solution temperature to facilitate balanced regeneration increases. Therefore the available flow temperature from the SOFC waste heat recovery circuit will have a significant influence on the concentration the desiccant solution can be operated at in the tri-generation system. Liquid desiccant air conditioning shows great potential as the thermally activated cooling technology in an SOFC tri-generation system, this is primary due to the effective use of low grade thermal energy. Regeneration of the $CHKO_2$ solution occurs from a temperature of 47.5 °C when operating the desiccant system at a 0.7 solution mass concentration.

Liquid desiccant air conditioning technology also shows promise for tri-generation integration, particularly with SOFC, due to its flexibility in operation. Operating the dehumidifier and regenerator independently or out of sync of one another has the potential to improve tri-generation system performance by (a) permitting the dehumidifier to operate at a moisture removal rate that is greater than the achievable regenerator moisture addition rate attainable with the instantaneous thermal input from the SOFC. The constant SOFC thermal output can be used to operate the regenerator for extended periods and make up the solution concentration difference, or (b) allowing the SOFC thermal output to be used in the day for building hot water demands and at night in the regenerator. In both these scenarios re-concentration of the desiccant solution can be considered a form of thermal energy storage, with a significant advantage of no losses over time. This operating concept is discussed in more detail in Sects. 6.3.4 and 7.4.2.

A numerical analysis of the complete desiccant air conditioning system, combining the dehumidifier and regenerator is presented in Sect. 4.2.2. Next, Sect. 3.6 presents the chapter conclusions.

3.6 Chapter Conclusions

This chapter has aimed to (1) assess a variety of liquid desiccant solutions with regard to their potential for dehumidification and suitability for integration into a liquid desiccant tri-generation system, and (2) develop mathematical models to evaluate the dehumidifier and regenerator membrane heat and mass exchangers.

A low cost, environmentally friendly potassium formate ($CHKO_2$) desiccant solution has been proposed as an appropriate working fluid in the liquid desiccant air conditioning system. Although the dehumidification performance is lower than the other solutions considered, other important attributes, such as: low density and viscosity, a lower regeneration temperature requirement, negligible environmental impact, low corrosiveness and cost make it a suitable choice. In the dehumidifier, the desiccant solution will be used in the concentration and temperature range of 0.65–0.7 °C and 20–30 °C respectively. This provides sufficient cooling in the proposed atmospheric conditions and has a regeneration temperature requirement in good thermal agreement with the SOFC.

Sections 2.3 and 2.5.4 of the literature review have presented theoretical and experimental performance assessments of the use of $CHKO_2$ in liquid desiccant air conditioning applications. It has been highlighted that there is limited published and available literature within the public domain regarding the specific physical characterisation of the $CHKO_2$ solution for liquid desiccant air conditioning applications. The work presented in Sect. 3.2 and Appendix A provides a contribution to knowledge with the assimilation of various sources of data to physically characterise the $CHKO_2$ solution in the required operating range for a liquid desiccant air conditioning application. Validation of the $CHKO_2$ dehumidifier/regenerator simulations with experimental results in Sect. 6.3 demonstrates the accuracy of the physical characterisation data presented.

The mathematical models and simulation process used to assess the dehumidifier and regenerator membrane heat and mass exchangers have been introduced. The models have been validated to within acceptable limits with published experimental data using a LiCl solution. Model validation with a $CHKO_2$ working solution is provided in Sect. 6.3. Parametric analyses for both the dehumidifier and regenerator have been carried out, focussing on the impact environmental and operating conditions have on performance. Conclusions have been drawn that are relevant to the integration of liquid desiccant air conditioning technology into an SOFC tri-generation system. Furthermore, the parametric analyses have strengthened the understanding of the operation, process and performance of liquid desiccant air conditioning; essential for the future theoretical and experimental work presented in the thesis.

Based upon the theoretical work carried out in this chapter, liquid desiccant air conditioning technology is an appropriate technological choice for a reliable, efficient and effective SOFC based tri-generation system, particularly on a domestic building scale. This is on account of four reasons:

(1) Shows effective cooling potential at low system capacities i.e. <10 kW.
(2) Effective use of low grade thermal energy—regeneration of the $CHKO_2$ solution occurs from a temperature of 47.5 °C when operating the desiccant system at a 0.7 solution mass concentration, demonstrating the potential for the development of an SOFC liquid desiccant tri-generation system.
(3) Operates at atmospheric pressure, with no harmful refrigerants or working fluids required.
(4) An operational advantage of the novel tri-generation system is that the dehumidification and regeneration process can operate independently of one another, thus increasing cooling capacity/maximising thermal energy utilisation from the SOFC. Regeneration of the desiccant solution using excess SOFC thermal output can be considered a form of no losses thermal energy storage.

Next, Chap. 4 presents a theoretical analysis of the complete tri-generation system, integrating the developed liquid desiccant dehumidifier and regenerator models with SOFC.

References

Çengel, Y.A., and A.J. Ghajar. 2011. *Heat and mass transfer fundamentals and applications*. New York: McGraw-Hill.

Conde, M. 2009. *Aqueous solutions of lithium and calcium chlorides: Property formulations for use in air conditioning equipment design*. M. C. Engineering.

Dwyer, T. 2014. *Liquid desiccants for dehumidification in building air conditioning systems*. The CIBSE Journal CPD Programme.

Enteria, N., and K. Mizutani. 2011. The role of the thermally activated desiccant cooling technologies in the issue of energy and environment. *Renewable and Sustainable Energy Reviews* 15(4): 2095–2122.

Ge, G., D. Ghadiri Moghaddam, A.H. Abdel-Salam, R.W. Besant, and C.J. Simonson. 2014. Comparison of experimental data and a model for heat and mass transfer performance of a liquid-to-air membrane energy exchanger (LAMEE) when used for air dehumidification and salt solution regeneration. *International Journal of Heat and Mass Transfer* 68: 119–131.

Jain, S., and P.K. Bansal. 2007. Performance analysis of liquid desiccant dehumidification systems. *International Journal of Refrigeration* 30(5): 861–872.

James, S. 1998. *A new working fluid 'Potassium Formate' for use in absorption heat pumps*. Masters of Philosophy, The University of Nottingham.

Jradi, M. and S. Riffat. 2014a. Energy performance of an innovative liquid desiccant dehumidification system with a counter-flow heat and mass exchanger using potassium formate. *Renewable Bioresources* 2(1).

Jradi, M. and S. Riffat. 2014b. Experimental and numerical investigation of a dew-point cooling system for thermal comfort in buildings. *Applied Energy* 132: 524–535.

Jradi, M. and S. Riffat. 2014c. Experimental investigation of a biomass-fuelled micro-scale tri-generation system with an organic Rankine cycle and liquid desiccant cooling unit. *Energy* 71: 80–93.

Liu, S. 2008. *A novel heat recovery/desiccant cooling system*. Ph.D., The University of Nottingham.

Lowenstein, A. 2008. Review of liquid desiccant technology for HVAC applications. *American Society of Heating, Refrigerating and Air-Conditioning Engineers* 14(6): 819–839.

Melinder, A. 2007. *Thermophysical properties of aqueous solutions used as secondary working fluids*. Ph.D. Thesis, KTH Energy and Environmental Technology.

Pilatowsky, I., R.J. Romero, C.A. Isaza, S.A. Gamboa, P.J. Sebastian, and W. Rivera. 2011. *Cogeneration fuel cell—Sorption air conditioning systems*. London: Springer.

Stephen, H., and T. Stephen. 1963. Solubilities of inorganic and organic compounds, volume 1. *Binary Systems Part* 1(1): 179.

Welty, J.R., C.E. Wicks, R.E. Wilson, and G.L. Rorrer. 2008. *Fundamentals of momentum, heat, and mass transfer*. Oregon: Wiley.

Chapter 4
SOFC Tri-generation System Modelling and Analysis

4.1 Introduction

In recent years, the dramatic increase in fossil fuel prices and the accompanying concerns regarding their environmental impact have driven governments, business and consumers towards cleaner energy resources and to use alternative methods for more efficient energy utilisation. Compared to a conventional separate generation scenario, energy conversion in a tri-generation configuration offers the potential for high energy efficiency and thus greater energy utilisation, providing a range of technical, environmental and economic benefits (Jradi and Riffat 2014d).

Chapter 3 has presented validated liquid desiccant dehumidifier and regenerator models. This chapter presents a theoretical analysis of the complete tri-generation system, integrating SOFC and liquid desiccant technology. The SOFC CHP system will produce an electrical and thermal output. The liquid desiccant dehumidifier will generate a cooling output. To facilitate continuous operation of the liquid desiccant air conditioning system the SOFC thermal output will be used for desiccant solution regeneration. As previously highlighted in Chap. 2, no work has been found in the literature regarding an SOFC liquid desiccant tri-generation system, thus the work in this chapter provides a contribution to knowledge through a theoretical analysis of the novel system. The aim of the chapter is to demonstrate the feasibility of combining SOFC and liquid desiccant air conditioning technology into an efficient and effective tri-generation system. The work presented seeks to quantify and qualify the benefits of combining the two technologies, to provide a comparison of performance compared to a conventional separated system and to offer a benchmark of performance/operational considerations for the experimental tri-generation system.

A schematic of the modelled tri-generation system is provided in Fig. 4.1, and is comprised of two main sub components, the SOFC CHP system and the desiccant air conditioning system. The complete tri-generation system is truly multi-functional, generating an electrical, heating and cooling output.

T. Elmer, *A Novel SOFC Tri-generation System for Building Applications*,
Springer Theses, DOI 10.1007/978-3-319-46966-9_4

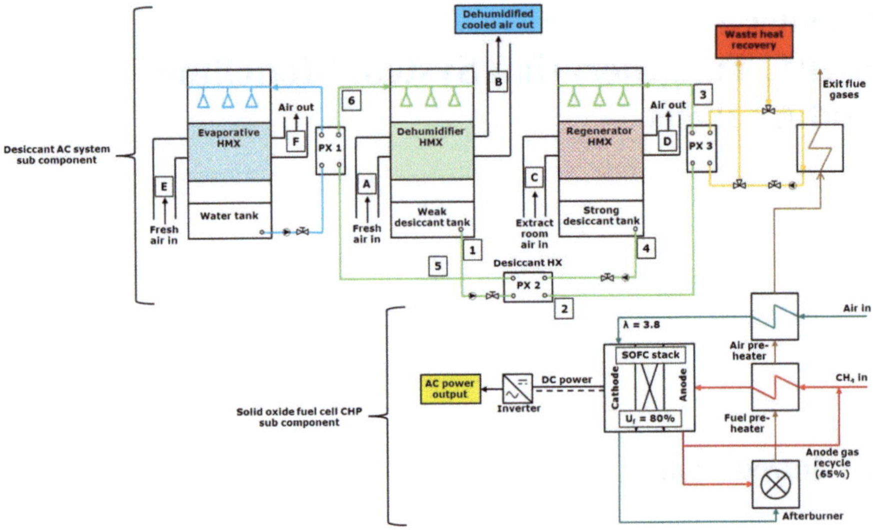

Fig. 4.1 The modelled tri-generation system

This chapter is split into two sections. Section 4.2 provides a theoretical performance analysis of the tri-generation system's SOFC and liquid desiccant sub-components. The sub-component assessment focuses on the impact changes in operating values have on their performance. The analysis serves two purposes (1) to improve the understanding of the sub-component operation and its influence on performance, and (2) selection of specific sub-component operating values to facilitate effective tri-generation system integration.

Section 4.3 integrates the SOFC CHP and liquid desiccant sub-components into a complete tri-generation system. Firstly, a specific tri-generation system case, based on defined selection criteria, is presented. Following this, a parametric analysis is used to investigate the effect changes in electrical and cooling capacities have on tri-generation system performance. A 24 h tri-generation system simulation is also provided. Finally, a climatic performance investigation is presented for three locations. Throughout, the performance of the novel SOFC tri-generation system is compared to a conventional separated system comprising grid electricity, gas fired boiler and vapour compression cooling. Finally, conclusions based on the theoretical assessment that are relevant to the experimental system are provided.

4.2 Tri-generation System Sub Component Analysis

In this section, the SOFC CHP and liquid desiccant air conditioning system sub components are described and their operational performance detailed through parametric analysis. Using the parametric analyses, the sub-component system sizes and operation are optimised to facilitate tri-generation integration. Next, Sect. 4.2.1 presents the SOFC CHP system sub-component description, modelling and parametric analysis.

4.2.1 Solid Oxide Fuel Cell CHP System

The aim of this section is to develop a SOFC CHP model so that the complete tri-generation system may be simulated in Sect. 4.3. The requirements of the SOFC CHP model are electrical and thermal output, electrical and CHP efficiency and heat to power ratio. The main independent variables considered in the SOFC modelling are: current density and fuel utilisation factor. These variables have been selected as they are influential on the performance of SOFC CHP and tri-generation systems (Al-Sulaiman et al. 2010).

A great advantage of SOFC technology is the potential for internally reforming a hydrocarbon fuel such as methane to hydrogen, thus removing the need for the expensive and complicated external reforming equipment seen in PEMFC systems. SOFCs also offer higher electrical efficiencies, long-term stability and relatively low costs compared to other fuel cell variants. However, disadvantages include high operating temperatures required for electrical and ionic conduction, meaning long start up times, issues with mechanical and chemical compatibility and little provision for on/off cycling. As a result, a large amount of research has focussed on lowering the operating temperatures of SOFCs to 500 °C and below (Tuyen and Fujita 2012; Fan et al. 2013). Furthermore, the maximum operating temperature of standard metals such as stainless steel is below 650 °C, therefore leading to much lower cost materials and components.

Unlike other fuel cells, SOFCs can have multiple cell geometries. There are three SOFC geometries: planar, co-planar and micro-tubular. In the most common planar type design, as shown in Fig. 4.2, an electrolyte is sandwiched between an anode and cathode to form flat cells, which are assembled into stacks. The fuel (hydrogen or short chain hydrocarbons) and oxidant (oxygen or air) flow though the unit via channels built in the anode and cathode respectively. A number of these cells are connected in series to form a stack to produce a useful electrical output. Conventionally, the anode, cathode and electrolyte materials used in SOFCs are: Ni-YSZ (Nickel-Yttria-stabalised zirconia), LSM (Lanthanum strontium manganite) and YSZ (Yttria-stabalised zirconia) respectively; however other material options do exist.

The YSZ electrolyte materials used in SOFCs do not become electrically and ionically active until they reach very high temperature and as a consequence the stacks have to run at temperatures ranging from 500 to 1000 °C. Air is supplied to the cathode, where reduction of oxygen into oxygen ions occurs: $O_2 + 4e^- \rightarrow 2O^{2-}$ The oxygen ions then diffuse through the solid oxide electrolyte to the anode where they electrochemically oxidise the hydrogen fuel to produce water and electrons: $2H_2 + 2O^{2-} \rightarrow H_2O + 4e^-$. An external circuit allows

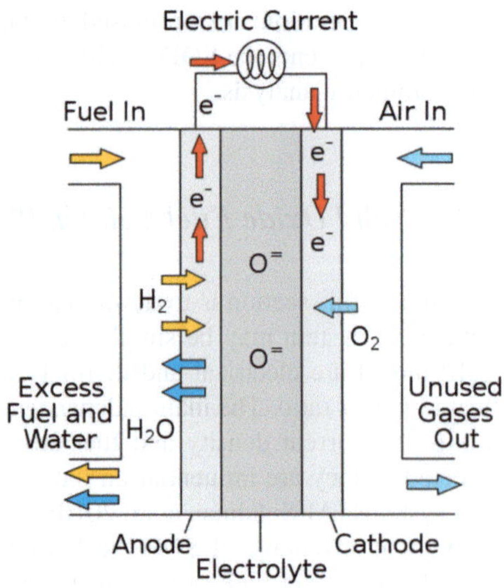

Fig. 4.2 Planar type SOFC construction and operation (Tuyen and Fujita 2012)

the flow of electrons from the anode to the cathode and thus current generation. Because there is a potential difference generated across the cell, electrical power is produced. The cycle then repeats as the released electrons re-enter the cathode. The overall reaction occurring in a SOFC is $2H_2 + O_2 \rightarrow 2H_2O$. When using methane as a fuel, the CH_4 reacts with O_2 via internal reforming, producing H_2O and some CO_2, as shown in Eq. 4.2. This is discussed in more detail below.

Due to its ready availability, natural gas is the most commonly used fuel in SOFC building applications. Because the typical composition of natural gas is 70–90 % methane, the modelling work assumes a direct methane fuel input to the SOFC from the domestic gas stream, an assumption adopted by previous researches in numerical studies (Blum et al. 2011). In reality the natural gas would need to be de-sulphurised with a small amount of pre-reforming before being supplied to the SOFC. When operating with methane, hydrogen is produced through internal steam reforming on the fuel cell anode. In this work, a planar type Ni-YSZ anode supported SOFC with LSM cathodes and thin YSZ electrolytes, operating with methane at a temperature of 800 °C has been modelled. A description of the SOFC modelling procedure is provided in Sect. 4.2.1.1.

4.2.1.1 SOFC Model Description

The modelling aims to determine the electrochemical and thermal performance of a methane fed internally reforming SOFC operating at 800 °C. The model is programmed using Engineering Equation Solver (EES).

In a fuel cell, it is the change in the Gibbs free energy of formation, $\overline{\Delta G_f}$, that gives the energy released during the electrochemical reaction in. This change is the difference between the Gibbs free energy of the products and the Gibbs free energy of the reactants, as shown in Eq. 4.1 (Larmine and Dicks 2003).

$$\overline{\Delta G_f} = \overline{\Delta G_f} \, products - \overline{\Delta G_f} \, reactants \tag{4.1}$$

The quantities of the products and reactants are usually considered in their per mole form $\bar{g}_f$ (J mol^{-1}). The overall reaction occurring in an SOFC operating on methane is shown in Eq. 4.2 (Pilatowsky et al. 2011).

$$CH_4 + 2O_2 \rightarrow 2H_2O + CO_2 \tag{4.2}$$

The product is two moles of H_2O and one mole of CO_2, the reactants are one mole of CH_4 and two moles of O_2. The change in Gibbs free energy per mole of substance is shown in Eq. 4.3.

$$\begin{aligned} \Delta\bar{g}_f &= \bar{g}_f \; of \; products - \bar{g}_f \; of \; reactants \\ &= \left((\bar{g}_f)_{H_2O} + \frac{1}{2} (\bar{g}_f)_{CO_2} \right) - \left(\frac{1}{2} (\bar{g}_f)_{CH_4} + (\bar{g}_f)_{O_2} \right) \end{aligned} \tag{4.3}$$

The Gibbs free energy of formation for the reaction shown in Eq. 4.3 is not constant; it changes with temperature and state. At a temperature of 800 °C, where the water product is in the steam phase, $\overline{\Delta g_f} = -818{,}400$ J mol^{-1} (Blomen 1993). The negative value denotes energy released. If the electrochemical process was reversible i.e. it were ideal with no losses to the environment, then all this Gibbs free energy would be converted to electricity, however in reality some energy is converted to heat.

The Gibbs free energy of formation $(\overline{\Delta g_f})$ can be used to find the reversible open circuit voltage (E_0) of the methane fed SOFC. This is shown in Eq. 4.4 (Tuyen and Fujita 2012).

$$E_0 = \frac{-\Delta\bar{g}_f}{zF} \tag{4.4}$$

F is the Faraday's constant 96,472.44 °C mol^{-1} and z denotes the number of electrons that pass through the external circuit for each mole of reactant. As shown in Eq. 4.5, for every mole of methane fed to the SOFC, eight electrons are released and circulate in the external circuit, thus $z = 8$ (Pilatowsky et al. 2011).

$$CH_4 + 4O^{2-} \rightarrow 2H_2O + CO_2 + 8e^- \tag{4.5}$$

Using Eq. 4.4, an SOFC operating at 800 °C will have a reversible open circuit voltage (E_0) of 1.06 V. However, E_0 assumes no irreversibilities. In practice the actual operating voltage (V_{cell}) is lower due to cell voltage drops. Some of the irreversibilities even apply when no current is drawn. The actual operating voltage of the SOFC (V_{cell}) is given in Eq. 4.6.

$$V_{cell} = E_{OCV} - \Delta V_{ohm} - \Delta V_{act} - \Delta V_{trans} \tag{4.6}$$

E_{OCV} is the non-reversible open circuit voltage of the cell, and considers the impact of fuel composition and ratio of fuel to oxidant and products. E_{OCV} is determined using the Nernst equation, shown in Eq. 4.7, and is a function of the partial pressures of the gases in the cell (Pilatowsky et al. 2011).

$$E_{OCV} = E_O + \frac{RT}{zF} ln\left(\frac{P_{CH_4}}{P_{H_2O} \cdot P_{CO_2}}\right) + \frac{RT}{zF} ln\left(P_{O_2}^2\right) \tag{4.7}$$

E_0 is the reversible open circuit voltage. R is the universal gas constant (8.314 J mol^{-1} K). P denotes the concentrations of the gas mixtures within the cell, given as partial pressures relative to atmospheric pressure (101,325 Pa). For the SOFC CHP system simulation, the partial pressures have been assumed constant, with $P_{CH_4} = 25,330$ and $P_{H_2O} = 75,990$ Pa, values taken from experimental data (Peters et al. 2002). Using these constants equates to a steam to carbon ratio (S/C) of 3:1. A minimum S/C ratio of two is required to achieve complete internal reforming of the methane fuel in the SOFC stack (Blum et al. 2011). As detailed below, the recycling of anode off-gas into the inlet fuel stream provides enough water for the internal reforming reaction to occur. Because the SOFC is fed by air, the partial pressure of the oxygen in the cell has been assumed constant at $P_{O_2} = 21{,}270$ Pa, as this is the partial pressure of oxygen in air. The partial pressure of the carbon dioxide product, $P_{CO_2} = 31{,}900$, is taken from the experimental work of Laosiripojana and Assabumrungrat (2007).
A description and quantification of the operating voltage losses are provided below.

Ohmic losses (ΔV_{ohm}) are caused due to resistance to the flow of electrons through the anode, cathode, electrolyte and various inter-connections. Equation 4.8 quantifies the Ohmic losses (Tuyen and Fujita 2012).

$$\Delta V_{ohm} = i.R_{ohm} \tag{4.8}$$

i is the current density in mA cm^{-2} of the cell, R_{ohm} is the electrical resistance of the cell in mΩ cm^{-2}. Tuyen and Fujita (2012) state that the largest contributor to Ohmic losses is from the transport resistance of O^{2-} in the electrolyte, and is strongly dependent upon temperature. The resistances of the other components can be considered constant due to their weak dependence on temperature and their small contribution to the total resistance. For a planar type SOFC with a YSZ electrolyte, Ni-YSZ anode and LSM cathode, the total electrical resistance can be calculated using Eq. 4.9 (Tuyen and Fujita 2012).

$$R_{ohm} = 0.0248e^{10100\left(\frac{1}{T} - \frac{1}{1273}\right)} + 0.0093 \tag{4.9}$$

T is the operating temperature of the cell, and has been set at 800 °C.

Activation losses (ΔV_{act}) are caused by the slowness of the reactions taking place on the surface of the electrodes, and are the sum of those occurring at the anode and cathode, shown in Eq. 4.10.

$$\Delta V_{act} = V_{act,anode} + V_{act,cathode} \tag{4.10}$$

The activation losses occurring at the electrodes, $V_{act,e}$ for both the anode and cathode, are determined using Eq. 4.11 (Al-Sulaiman et al. 2010).

$$V_{act,e} = \frac{R.T}{F}\left(\sin h^{-1}\left(\frac{i}{2.io_e}\right)\right) \tag{4.11}$$

io_e is the exchange current density of the electrodes. In this model, 0.65 A cm^{-2} for the anode and 0.25 A cm^{-2} for the cathode have been selected, a reasonable assumption for a planar SOFC with Ni-YSZ anodes and LSM cathodes (Al-Sulaiman et al. 2010).

Transmission losses (ΔV_{trans}) are caused due to the change in concentration of the reactants at the surface of the electrodes as the fuel is used. Equation 4.12 quantifies the transmission losses.

$$\Delta V_{trans} = m\,exp(n.i) \tag{4.12}$$

Equation 4.12 is an empirical correlation. Tuyen and Fujita (2012) state that it provides a robust and accurate prediction of the transmission losses in a SOFC. m and n are constants, 3.1×10^{-5} V and 8.1×10^{-3} mA cm^{-2} have been selected respectively (Tuyen and Fujita 2012).

Once the cell operating voltage is known, the DC power output in Watts, from the SOFC stack is calculated using Eq. 4.13.

$$\dot{W}_{elec,DC} = I \cdot V_{cell} \cdot \mathrm{n}_{cell} \tag{4.13}$$

I is the current in Amps of one cell and is determined by multiplying the cell area by the operational current density. n_{cell} is the number of cells in the stack, and is selected based on the power output requirements. If the cells are connected in series the voltage of the stack is the sum of all the cell voltages. The operating cell efficiency is calculated using Eq. 4.14.

$$\eta_{cell} = \mathrm{u_f}\frac{V_{cell}}{\dot{LHV}_{eV,CH_4}}100\,\% \tag{4.14}$$

$\mathrm{u_f}$ is the fuel utilisation factor of the cell, and has been assumed as 80 % for the current model, a reasonable assumption for a planar type SOFC (Al-Sulaiman et al. 2010; Blum et al. 2011). To calculate the cell efficiency with respect to the SOFC operating voltage, the heating value of the methane fuel needs to be given in eV. For a SOFC operating at 800 °C, the H_2O product will be in the vapour phase, thus the lower heating value (LHV_{eV}) is used, for methane this is 1.039 eV (Blomen et al. 1993). Once the SOFC DC power output and cell efficiency is known, the total energy input in Watts to the SOFC ($\dot{Q}_{CH_4}$) is calculated using Eq. 4.15.

$$\dot{Q}_{CH_4} = \frac{\dot{W}_{elec,DC}}{\eta_{cell}} \tag{4.15}$$

The mass flow rate of the methane fuel into the SOFC stack is determined using Eq. 4.16. The *LHV* of CH_4 is given in J $kg^{-1} = 5 \times 10^7$

$$\dot{m}_{CH_4} = \frac{\dot{Q}_{CH_4}}{L\dot{H}V_{CH_4}} \quad (4.16)$$

The net stack AC power output in Watts is determined using Eq. 4.17.

$$\dot{W}_{elec,AC} = \left(\dot{W}_{elec,DC} - \dot{W}_{BoP}\right) \cdot \eta_{inverter} \quad (4.17)$$

η_{inverter} is the inverter efficiency, and has been assumed to be 90 %, a reasonable assumption for AC power production in small scale fuel cell systems (Blum et al. 2011). $\dot{W}_{\mathrm{BoP}}$ is the electrical consumption in Watts of the balance of plant (BoP) for components such as fans, pumps and control devices. Blum et al. (2011) states that the BoP power consumption is highly dependent upon the system, but based on empirical experience, the BoP will consume 5–10 % of the gross DC output from the SOFC stack. In this model $\dot{W}_{BoP}$ has been assumed as 10 %.

Heat up of an SOFC stack to operating temperature is usually achieved through the combustion of fuel (methane) in the afterburner. In this modelling work, only instantaneous SOFC CHP system performance is investigated and thus it is assumed that the SOFC system is already up to temperature. Heat-up energy input is not considered. Because of the high operating temperatures of SOFCs and the exothermic nature of the electrochemical reactions, a large quantity of heat is produced whilst generating electricity. In order to improve the overall efficiency of the SOFC system, this heat should be recovered. The thermal energy output and resulting waste heat recovery $\left(\dot{Q}_{\mathrm{WHR}}\right)$ from the SOFC system is derived from the hot gases leaving the SOFC stack (temperature assumed to be equal to the SOFC operating temperature) plus the heat of combustion of the unconsumed hydrogen fuel leaving the anode and flowing to the afterburner (Blum et al. 2011; Tuyen and Fujita 2012).

Anode gas recycling means that a portion of the unconsumed fuel leaving the SOFC stack, which is a function of the fuel utilisation factor, is re-circulated back to the inlet gas stream, and the rest is passed on to the afterburner. The anode gas recycle ratio has been set at 65 % as this provides enough water in the inlet fuel stream for internal reforming to occur (Blum et al. 2011). Although anode gas recycling reduces the amount of fuel entering the afterburner and thus the resulting heat output of the SOFC system, it does simplify the entire system due to the omission of steam generation components. For the modelled SOFC system, the mass ratio of air to fuel (λ) entering the fuel cell stack is 3.8. In such a design, this ratio provides enough air for the reactions and sufficient stack cooling.

The hot gases leaving the afterburner are used to pre-heat the inlet fuel and air streams to the required cell temperature, and finally transfer heat to the waste heat recovery (WHR) circuit. The recoverable thermal energy from the SOFC system $\left(\dot{Q}_{\mathrm{waste,SOFC}}\right)$ in Watts is shown in Eq. 4.18.

$$\dot{Q}_{waste,SOFC} = \left(\dot{Q}_{stack} + \dot{Q}_{afterburner}\right) - \left(\dot{Q}_{fuel} + \dot{Q}_{air}\right) \quad (4.18)$$

This recoverable thermal energy is used to heat water in the WHR circuit, which is then either used for direct heating applications or to regenerate desiccant solution depending if the system is operating in a CHP or tri-generation configuration. All heat exchanger processes have been evaluated using the ε method, with ε_{PX} assumed as 80 %. The net recoverable heat from the SOFC CHP system $(\dot{Q}_{WHR})$ is determined using Eq. 4.19.

$$\dot{Q}_{WHR} = \dot{m}_w c_{p,w} \left(T_{WHR,flow} - T_{WHR,return} \right) \tag{4.19}$$

$\dot{m}_{water}$ is the mass flow rate of water in the WHR circuit, and has been assumed constant at 0.042 kg s^{-1}. c_p is the specific heat capacity of the water in J kg^{-1} K, and is a function of temperature. $T_{WHR,flow}$ and $T_{WHR,return}$ are the respective flow and return temperatures of the water in the WHR circuit, given in °C. During simulation $T_{WHR,return} = 45$ °C. The SOFC CHP system sub-component schematic in Fig. 4.1 shows the WHR circuit and anode gas recycling concepts described above. The net electrical efficiency of the SOFC CHP system is related to the net AC electrical output, and is calculated using Eq. 4.20.

$$\eta_{elec} = \frac{\dot{W}_{elec,AC}}{\dot{Q}_{CH_4}} \tag{4.20}$$

The combined heat and power or co-generation heating efficiency is the sum of the electrical and thermal outputs, and is shown in Eq. 4.21.

$$\eta_{CHP} = \frac{\dot{W}_{elec,AC} + \dot{Q}_{WHR}}{\dot{Q}_{CH_4}} \tag{4.21}$$

An important parameter in CHP systems is the heat to power ratio (HP) which indicates the proportion of generated heat to electrical power. Equation 4.22 is used to determine the SOFC CHP systems HP. For a planar type SOFC CHP system of less than 10 kW_e, the HP will be between 0.5 and 1 (Elmer et al. 2015b).

$$HP = \frac{\dot{W}_{elec,AC}}{\dot{Q}_{WHR}} \tag{4.22}$$

Liu and Barnett (2003) provide experimental electrochemical performance data for an internally reforming SOFC with thin YSZ electrolytes on porous Ni-YSZ anodes, operating on methane at a cell temperature of 800 °C. Validation of the developed SOFC model with the experimental data is provided in Fig. 4.3. Over a current density range of 0–1000 mA cm^{-2} the developed SOFC model shows good agreement with the published experimental data.

The SOFC model has been validated, to a suitable degree, with published experimental data. The SOFC model can therefore be used with confidence in the SOFC parametric and tri-generation system analysis.

Next, Sect. 4.2.1.2 presents the SOFC CHP system parametric analysis.

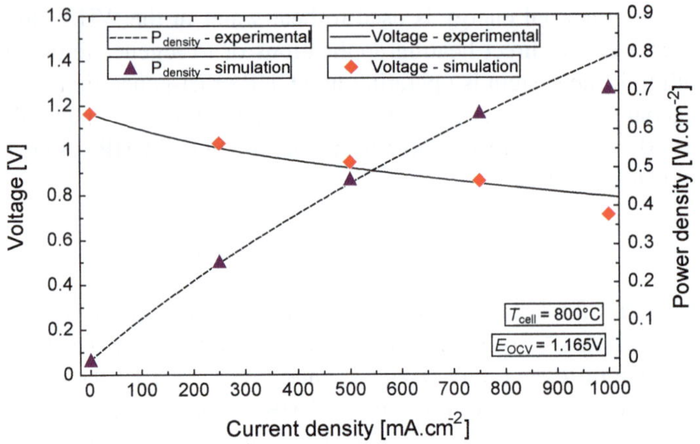

Fig. 4.3 SOFC electrochemical validation

4.2.1.2 SOFC CHP System Parametric Analysis

This section presents a SOFC CHP system parametric analysis along with a specific case based on defined selection criteria. To simplify the modelling process, some assumptions have been made, and are listed below.

- All processes are considered steady-state
- The fuel is methane (CH_4) and inlet temperature of the gases is 20 °C (ambient)
- A cell temperature of 800 °C has been chosen as this will achieve a methane conversion rate of >99 %
- The system operates at atmospheric pressure, with no pressure loss. The system is assumed to be adiabatic i.e. no heat loss
- All efficiencies are related to the *LHV* of the fuel

Next, a parametric analysis is presented to demonstrate the impact SOFC current density and fuel utilisation factor have on SOFC CHP system performance. Specific consideration to the tri-generation system is made throughout.

Figure 4.4a presents the impact current density, in the range of 0–1200 mA cm^{-2}, has on the operating cell voltage and power density. It is evident that the highest voltage is 1.036 V and achieved when no current is drawn i.e. E_{OCV}. Note this is lower than the reversible open circuit voltage (E_O) of 1.06 V. Initially there is a very small fall in voltage with current density, then the reduction between 100 and 900 mA cm^{-2} is fairly linear. The voltage drop above 900 mA cm^{-2} is much greater. As demonstrated in Fig. 4.4b the activation losses are responsible for the initial fall in voltage. SOFCs do not suffer so much from activation losses compared to PEMFCs, where the initial drop is much more pronounced. Following this initial drop, the fairly linear decline in voltage is primarily due to Ohmic losses i.e. electrical resistance of the electrodes, and the

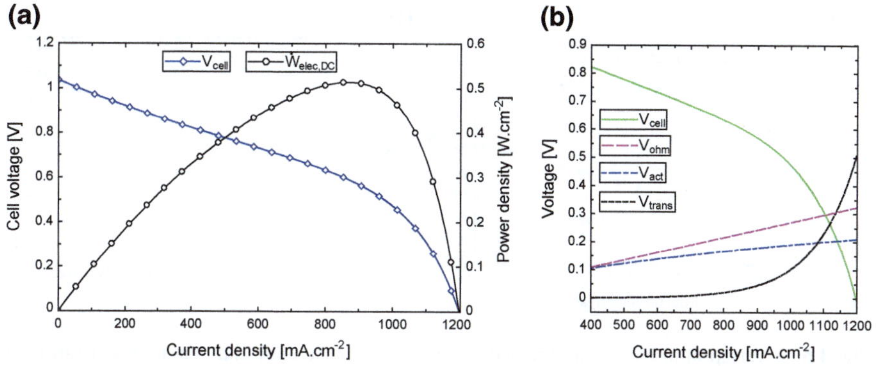

Fig. 4.4 **a** Variation of cell voltage and power density with current density, and **b** a break-down of the voltage losses with current density

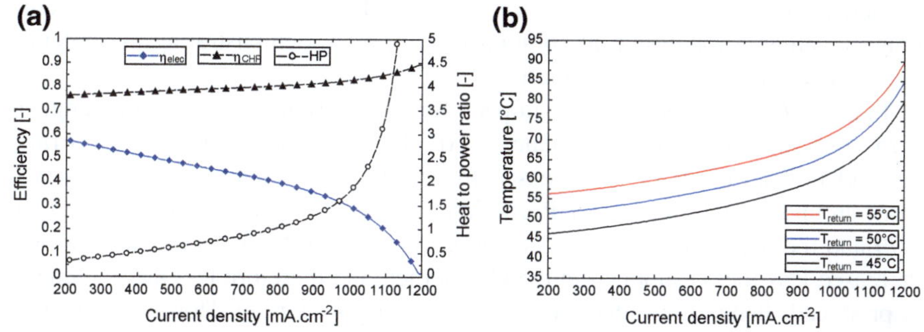

Fig. 4.5 **a** Variation in electrical efficiency, cogeneration efficiency and heat to power ratio with current density, and **b** the variation in WHR outlet water temperature with current density and return water temperature

resistance to the flow of ions in the electrolyte. Finally, the large drop in voltage is attributed to transmission losses. At high current densities there is a depletion of reactants at the electrodes, causing a rapid loss of voltage. Figure 4.4a demonstrates a peak power density of 0.5 W cm^{-2} is achieved at 900 mA cm^{-2}. In reality SOFC, operational current densities are between 400 and 800 mA cm^{-2}, giving an operating cell voltage range of 0.63–0.82 V (Larmine and Dicks 2003).

Figure 4.5a demonstrates the impact current density, in the range of 200–1200 mA cm^{-2}, has on electrical efficiency, CHP efficiency and the heat to power ratio. As current density is increased the electrical efficiency decreases from around 0.57–0.3, the CHP efficiency increases from 0.76 to 0.83 and the heat to power ratio increases from 0.33 to 1.8. The current density of the SOFC needs to be at 625 mA cm^{-2} or less to achieve an electrical efficiency of 45 % or greater. Figure 4.5b demonstrates that as current density is increased, the outlet water

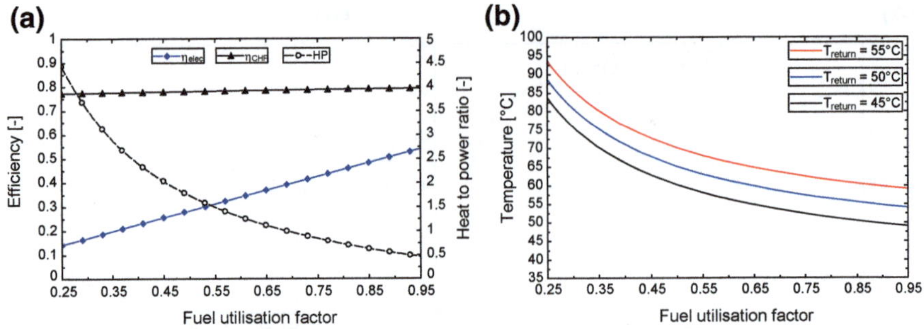

Fig. 4.6 **a** Variation in electrical efficiency, cogeneration efficiency and heat to power ratio with fuel utilisation factor, and **b** the variation in WHR outlet water temperature with fuel utilisation factor and return water temperature

temperature in the WHR circuit increases. This is because the heat to power ratio is increasing. The outlet water temperature in the 400–800 mA cm^{-2} operating range is suitable for desiccant solution regeneration, demonstrating the potential for tri-generation system integration.

Figure 4.6a demonstrates the impact fuel utilisation factor, in the range of 0.25–0.95, has on electrical efficiency, CHP efficiency and the heat to power ratio. As fuel utilisation factor is increased the electrical efficiency increases linearly from around 0.14–0.54, the CHP efficiency increases very slightly from 0.77 to 0.79 and the heat to power ratio decreases from 4.42 to 0.46. As fuel utilisation is increased less fuel is supplied to the afterburner, and therefore less thermal energy is produced, hence the reduction in the heat to power ratio. However, electrical efficiency is a function of fuel utilisation, and thus the CHP efficiency remains largely unaffected. The fuel utilisation factor needs to be at 0.785 or more to achieve an electrical efficiency of 45 % or greater. Figure 4.6b demonstrates that as fuel utilisation factor is increased, the outlet water temperature in the WHR circuit reduces. This is because the heat to power ratio is decreasing. In most SOFC systems, the fuel utilisation factor is in the range of 0.6–0.85, and dependent upon the particular stack design. Both the electrical efficiency and outlet water temperature targets in this fuel utilisation factor range are suitable for the tri-generation system.

Figures 4.4, 4.5 and 4.6 demonstrate the impact changes in operating values have on the performance of the SOFC CHP system. Table 4.1 presents a specific case for the SOFC CHP system based on the following selection criteria: (1) realistic operating values found in SOFC CHP systems of this capacity and (2) achieving an electrical output of 1.5 kW_e at a net electrical efficiency of 45 % or greater, as set out in the technical objectives in Chap. 1.

A single cell area of 81 cm^2 (9 cm × 9 cm) has been selected. This is based upon a commonly used cell size in SOFCs (Al-Sulaiman et al. 2010; Tuyen and Fujita 2012). To achieve an electrical efficiency of 45 % or more, a current density of 600 mA cm^{-2} has been selected. This is a reasonable assumption for a planar

Table 4.1 Input data for specific SOFC CHP system case

Parameter	Value
Single cell area (cm^2)	81
Baseline cell current density ($mA\ cm^{-2}$)	600
Number of cells in stack	53
Fuel utilisation factor (%)	80
Oxygen utilisation factor (%)	30
Anode off-gas recycle factor (%)	65
Heat exchanger effectiveness (%)	80
Balance of plant consumption (%)	10
Inverter efficiency (%)	90
Mass ratio of air to fuel (λ)	3.8
$T_{WHR,return}$ (°C)	45
WHR water volumetric flow ($L\ min^{-1}$)	2.5

Table 4.2 SOFC CHP system model performance data

Parameter	Value
V_{cell} (V)	0.7316
$\dot{W}_{elec,AC}$ (W)	1526
$\dot{Q}_{WHR}$ (W)	1116
$\dot{Q}_{CH_4}$ (W)	3345
η_{elec} (%)	45.63
$\eta_{thermal}$ (%)	33.37
η_{CHP} (%)	78.98
HP	0.73
$T_{WHR,flow}$ (°C)	51.41

type anode supported SOFC with YSZ electrolytes, Ni-YSZ anodes and LSM cathodes (Larmine and Dicks 2003). To achieve a 1.5 kWe output 53 cells are required. Based on the selected input values shown in Table 4.1, the performance of the SOFC CHP system is presented in Table 4.2.

The selected input values for the SOFC CHP system shown in Table 4.1 are both realistic and fulfil the technical objectives of the thesis. As a result, these will be used in the SOFC tri-generation system modelling in Sect. 4.3. The SOFC CHP system has an electrical and thermal output of 1526 and 1116 W respectively, with an electrical efficiency of 45.63 %. The outlet WHR water temperature in the selected case is sufficient to facilitate desiccant solution regeneration at a 0.65 solution mass concentration, as demonstrated in Chap. 3.

Section 4.2.1 has presented a theoretical component analysis of a validated SOFC CHP system model. A model description, parametric analysis and specific case based on defined selection criteria have been presented. The parametric analysis has served to demonstrate the relationship between current density and fuel utilisation factor on SOFC CHP system performance. The specific SOFC CHP case demonstrates that SOFC and liquid desiccant air conditioning technology are suitable for tri-generation system integration. Next, Sect. 4.2.2 presents the liquid desiccant air conditioning system sub-component description and parametric analysis.

4.2.2 Liquid Desiccant Air Conditioning System

The individual validated desiccant dehumidifier and regenerator models have been evaluated in detail in Chap. 3, therefore only the incorporation of the dehumidifier and regenerator into a complete desiccant air conditioning system are provided in this section. Figure 4.1 shows a schematic of the entire tri-generation system, including the modelled desiccant air conditioning sub component. In order for the desiccant air conditioning system to operate continuously, the mass of vapour absorbed by the desiccant solution in the dehumidifier must be removed in the regenerator, as demonstrated in Eqs. 4.23 and 4.24. Thermal input to the desiccant air conditioning system is through the heating of the desiccant solution to a temperature such that the mass balances are satisfied.

$$dm_{a,deh} = -dm_{a,reg} \tag{4.23}$$

$$\Delta X_{sol,deh} = -\Delta X_{sol,reg} \tag{4.24}$$

To form a complete desiccant air conditioning system the dehumidifier and regenerator are combined, alongside an evaporative cooling tower, desiccant to desiccant heat exchanger and a regenerative heat source. The cooling tower is included to cool the liquid desiccant solution prior to dehumidification. The desiccant to desiccant heat exchanger is used to pre-cool and pre-heat the desiccant solution as it flows between the dehumidifier and regenerator respectively. The regenerative heat source is used to raise the temperature of the descant solution to allow de-sorption of the moisture added in the dehumidifier. In the tri-generation system model, the thermal energy input comes from the SOFC CHP system.

Chapter 3 provides the equations used to calculate the state points of the air (Eqs. 3.18 and 3.20) and the desiccant solution (Eqs. 3.19 and 3.21) for both the dehumidifier and regenerator. These equations will not be presented again; however, the desiccant air conditioning process in a tri-generation system context is described below. The system operation is described with reference to the numbered desiccant flows and lettered air flows shown in Fig. 4.1.

During dehumidification and cooling of the fresh air (process A–B), strong cool desiccant solution (point 6) is sprayed downwards through the dehumidifier membrane HMX. The desiccant solution absorbs water and becomes weak (point 1). In order for the solution to have its dehumidification capacity restored it needs to be regenerated. First, the desiccant solution is pre-heated by hot desiccant solution flowing from the regenerator in the desiccant to desiccant heat exchanger (PX2).

The desiccant solution outlet temperature (point 2) is calculated using Eq. 4.25. The concentration remains unchanged. C denotes the heat capacity rate in J K^{-1}.

$$T_{sol,2} = T_{sol,1} + \left[\frac{\left(C_{min}\left(T_{sol,4} - T_{sol,1}\right)\right)\varepsilon_{PX}}{C_{sol,1}} \right] \quad (4.25)$$

The weak warm desiccant solution (point 2) is then heated to a temperature ($T_{sol,3}$) such that the mass balances in Eqs. 4.23 and 4.24 are satisfied (point 3). The thermal energy input at PX3 needed to raise the desiccant solution to the required temperature is determined using Eq. 4.26.

$$\dot{Q}_{reg} = \left(\dot{m}_{sol} c_{p_{sol}} \left(T_{sol,3} - T_{sol,2}\right)\right) / \varepsilon_{PX} \quad (4.26)$$

During regeneration, the weak hot desiccant solution is sprayed downward through the regenerator membrane HMX, and the moisture absorbed by the desiccant solution in the dehumidifier is desorbed into the air stream (process C–D). Following regeneration, the desiccant solution is strong and hot (point 4). Before it can be used again for dehumidification, it needs to be cooled. Initially the strong hot desiccant solution is pre-cooled by the cool desiccant solution flowing from the dehumidifier in the desiccant to desiccant heat exchanger (PX2). The desiccant solution outlet temperature (point 5) is calculated using Eq. 4.27. The solution concentration remains unchanged.

$$T_{sol,5} = T_{sol,4} - \left[\frac{\left(C_{min}\left(T_{sol,4} - T_{sol,1}\right)\right)\varepsilon_{PX}}{C_{sol,4}} \right] \quad (4.27)$$

Following pre-cooling of the desiccant solution, it is further cooled in PX1 with water flowing from an evaporative cooling tower (process E–F). The cooling tower outlet water temperature ($T_{cw,out}$) is calculated based on wet bulb effectiveness (η_{wb}) using Eq. 4.28. To replicate realistic operating conditions, the cooling tower inlet water temperature ($T_{cw,in}$) and the air wet-bulb temperature ($T_{wb,a}$) are assumed equal to the ambient environment. η_{wb} is assumed constant at 75 %.

$$T_{cw,out} = T_{cw,in} - \left(\eta_{wb}\left(T_{cw,in} - T_{wb,a}\right)\right) \quad (4.28)$$

The desiccant solution outlet temperature (point 6) is then calculated using Eq. 4.29. The solution concentration remains unchanged.

$$T_{sol,6} = T_{sol,5} - \left[\frac{\left(C_{min}\left(T_{sol,5} - T_{cw,out}\right)\right)\varepsilon_{PX}}{C_{sol,5}} \right] \quad (4.29)$$

Following solution cooling, the strong cool desiccant solution flows back to the dehumidifier and the process begins again. During tri-generation system modelling the thermal energy output from the SOFC CHP system $\left(\dot{Q}_{WHR}\right)$ must be greater than or equal to the thermal energy requirement of the regeneration process $\left(\dot{Q}_{reg}\right)$ for the tri-generation system process to be possible. As highlighted in Sect. 4.1, the primary consideration for the selection of sub-component operating values is that the SOFC thermal output matches the thermal requirement of the liquid desiccant system.

As highlighted in Sect. 3.5.2, regenerator capacity increases as the temperature and relative humidity of the inlet air stream is reduced. Thus, for both the liquid desiccant air conditioning system and tri-generation system analysis presented in Sects. 4.2.2.1 and 4.3 respectively, it is proposed that the dehumidifier and evaporative cooler operate on fresh outside air, and the regenerator operates on extracted room air, with an assumed condition of 26 °C and 60 % RH, an appropriate summer operative condition for a domestic dwelling according to CIBSE Guide A (CIBSE 2006).

Next, the liquid desiccant air conditioning system parametric analysis is presented.

4.2.2.1 Liquid Desiccant Air Conditioning System Parametric Analysis

This section presents the results from the liquid desiccant air conditioning system parametric performance analysis. The performance has been analysed with respect to: inlet air relative humidity, inlet air temperature, desiccant solution mass concentration and desiccant solution volumetric flow. A 24 h simulation of the desiccant system operating in a UK summer climate is also presented. This is to demonstrate the dynamic changes in the system over a daily cycle. The performance metrics used to evaluate the desiccant system are thermal coefficient of performance (COP_{th}) shown in Eq. 4.30, cooling output, required regenerator solution temperature and regenerator thermal input.

$$\mathrm{COP}_{th} = \frac{\dot{Q}_{cooling}}{\dot{Q}_{reg}} \tag{4.30}$$

The electrical COP (COP_{el}) of the system is calculated using Eq. 4.32. $\dot{W}_{aux,des}$ is the liquid desiccant system's electrical requirement (fans and pumps).

$$\mathrm{COP}_{el} = \frac{\dot{Q}_{cooling}}{\dot{W}_{aux,des}} \tag{4.31}$$

For a tri-generation system application, a higher COP_{th} is desirable as it means greater utilisation of thermal energy and thus improved tri-generation system efficiency. Table 4.3 provides the baseline input values for the desiccant air conditioning system parametric analysis. The analysis employs a potassium formate ($CHKO_2$) desiccant solution. The geometry, desiccant solution volumetric flow and air volumetric flow are the values used in Chap. 3 (Tables 3.10 and 3.11), which are based on the experimental SDCS presented in Chap. 6, and are thus realistic operating values which can be used with confidence.

Table 4.3 Baseline input values for desiccant system parametric analysis

Parameter	Value
Length of HMX exchanger (mm)	410
Height of HMX (mm)	210
Width of air channel (mm)	10
Inlet air flow ($m^3\ h^{-1}$)	191
Dehumidifier air temperature (°C)	30
Dehumidifier air relative humidity (%)	70
Regenerator air temperature (°C)	26
Regenerator air relative humidity (%)	60
Inlet desiccant flow ($L\ min^{-1}$)	3.2
Desiccant solution mass concentration	0.7
Cooling tower wet bulb effectiveness (%)	75
Plate heat exchanger effectiveness (%)	80

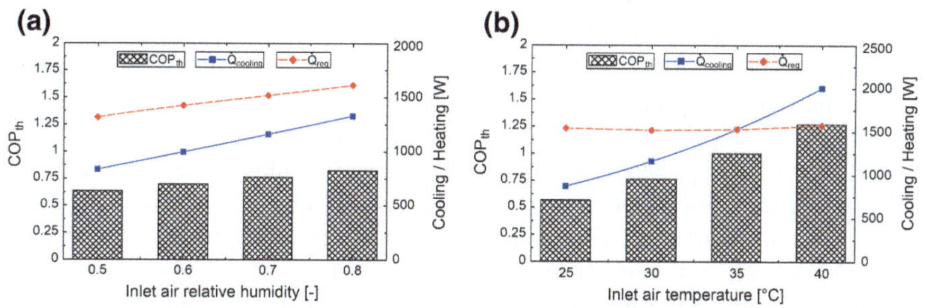

Fig. 4.7 Variation in liquid desiccant air conditioning system performance with **a** inlet air relative humidity, and **b** inlet air temperature

The desiccant solution volumetric flow in the dehumidifier is equal to the regenerator, as is the volumetric air flow. However, as discussed in Sect. 4.2.2, the inlet air to the dehumidifier is fresh outside air, and the inlet air to the regenerator is extracted room air. The inlet desiccant solution temperature is calculated based on the cooling water temperature, which is a function of the inlet air (fresh outside) wet bulb temperature.

Figure 4.7a demonstrates that as the inlet air relative humidity to the dehumidifier increases so does the cooling output and the required regenerator thermal input. Because the increase in cooling output is greater than the increase in regenerator thermal input, the COP_{th} increases with inlet air relative humidity, from 0.64 at 50 % RH to 0.82 at 80 % RH. As the inlet air relative humidity increases so does the required regeneration solution temperature from 44.75 °C at 50 % RH to 50.25 °C at 80 % RH.

Figure 4.7b demonstrates that as the inlet air temperature to the dehumidifier increases so does the cooling output, however the required regenerator thermal

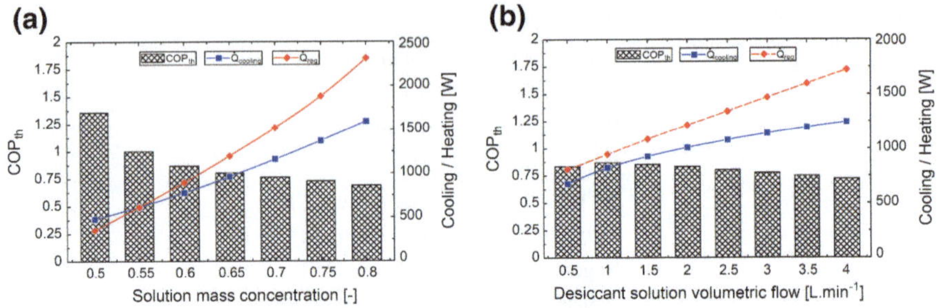

Fig. 4.8 Variation in liquid desiccant air conditioning system performance with **a** inlet solution mass concentration, and **b** desiccant solution volumetric flow

input remains almost constant. As a result the COP_{th} of the system increases dramatically, from 0.57 at 25 °C to 1.27 at 40 °C. As the inlet air temperature increases so does the required regeneration solution temperature from 46.25 °C at 25 °C to 54.5 °C at 40 °C. The regenerator thermal input remains constant even though the required solution temperature increases. This is because the ambient environmental temperature, which the stored solution is equal to, is higher.

For a particular application and desired outlet air condition the desiccant solution mass concentration cannot be randomly changed. However, the data presented in Fig. 4.8a is provided for comparative purposes, demonstrating how changes in the working desiccant solution mass concentration impacts desiccant system performance. Figure 4.8a demonstrates that as the working desiccant solution mass concentration increases so does the cooling output and required regenerator thermal input. Because the increase in cooling output is less than the increase in regenerator thermal input, the COP_{th} decreases, from 1.357 at a 0.5 solution mass concentration to 0.69 at a 0.8 solution mass concentration. Although a higher COP_{th} is desirable especially in a tri-generation system, a meaningful cooling/dehumidification output is also desirable and thus the lowest solution mass concentration shouldn't necessarily be used. With a thermal input of around 1200 W from the SOFC CHP system, a solution mass concentration between 0.6 and 0.65 should be selected. As the operating solution mass concentration increases, so does the required regeneration solution temperature from 32 °C at a 0.5 solution mass concentration to 59.25 °C at a 0.8 solution mass concentration. At a solution mass concentration between 0.6 and 0.65, the required regeneration solution temperature is 39–44 °C. It can be concluded that a potassium formate solution is an appropriate choice of working fluid in the liquid desiccant air conditioning system due to its low temperature regeneration requirement, which is in good thermal agreement with the SOFC CHP systems outlet water temperature.

Figure 4.8b demonstrates that as the desiccant solution volumetric flow in the system (dehumidifier and regenerator) is increased, both the cooling output, and required regenerator thermal input increases. The COP_{th} increases between 0.5 and 1 L min^{-1}, however, between 1 and 4 L min^{-1}, the COP_{th} decreases from 0.87

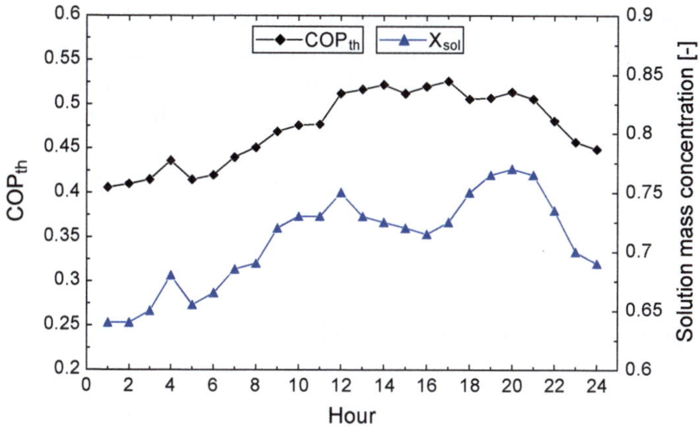

Fig. 4.9 24 h simulation of the liquid desiccant air conditioning system operating in a UK climate

to 0.72. This is primarily due to a sharp increase in required regenerator thermal input as a result of more desiccant solution flowing in the heating circuit.

Using Nottingham summer climatic data (temperature and relative humidity) a dynamic simulation of the liquid desiccant air conditioning system operating over a 24 h time period is presented in Fig. 4.9. The operating values used for the desiccant system are those provided in Table 4.3. The requirement of a constant supply air condition of around 20 °C and 54 % relative humidity ($\omega_{a,in} = 0.0078$ kg_{vapour}/kg_{dryair}) was used during the assessment. In order to maintain the desired supply air condition the solution mass concentration needs to be increased as the dehumidifier inlet air absolute humidity increases. Over the 24 h period the potassium formate solution mass concentration ranges from 0.64 to 0.77 and the COP_{th} ranges between 0.41 and 0.53, with an average of 0.48. As the inlet air absolute humidity increases, the desiccant solution mass concentration needs to be greater and thus more thermal input is required. This results in poorer system (COP_{th}) performance.

Section 4.2.2 has presented a theoretical component analysis of the liquid desiccant air conditioning system. A model description, parametric analysis and 24 h dynamic simulation have been provided. The parametric analysis has served to demonstrate the impact inlet air relative humidity, inlet air temperature, desiccant solution mass concentration and desiccant solution volumetric flow have on desiccant system performance. The results demonstrate effective use of low grade thermal energy (<60 °C). Whilst remaining within the boundaries of realistic operating values and a regenerator solution temperature of 40–50 °C, COP_{th} of 0.8 are attainable, demonstrating the potential for effective tri-generation system integration. To facilitate successful tri-generation system integration, the liquid desiccant regenerative thermal requirement needs to be less than or equal to the SOFC CHP thermal output. The knowledge and experience gained through the liquid desiccant modelling presented in Chap. 3 and Sect. 4.2.2 has been used to inform the design and operation of the liquid desiccant units presented in Chaps. 5 and 6. This has primarily focussed on discussions with the manufacturer ISAW regarding the size of the HMX and the provision for thermal energy input from an external heating source.

Next, Sect. 4.3 presents the simulation results from the complete tri-generation system analysis.

4.3 Tri-generation System Analysis

In this section, the validated SOFC CHP and liquid desiccant air conditioning sub-component models are combined to form a complete SOFC liquid desiccant tri-generation system model. Figure 4.1 presents a schematic of the novel system. The SOFC CHP system generates heat and power simultaneously. In CHP mode the heat is used directly in heating applications. In tri-generation mode, the heat is used to regenerate the desiccant solution so that the desiccant air conditioning system can operate continuously and dehumidified cool air can be produced. The performance of the tri-generation system will be assessed based on its tri-generation efficiency, as defined in Eq. 4.32.

$$\eta_{tri} = \frac{\dot{W}_{elec,net} + \dot{Q}_{WHR,net} + \dot{Q}_{cooling}}{\dot{Q}_{\mathrm{CH_4}}} \tag{4.32}$$

The tri-generation system efficiency is defined as the ratio of the overall tri-generation system energy conversion (electricity + heating + cooling) over the total amount of energy input to the system. The net electrical output is defined as the total AC electrical output from the SOFC less the electrical requirement to operate the liquid desiccant systems pumps and fans, and has been assumed constant, from empirical data, at 110 W. The net heat output is defined as the total thermal output from the SOFC CHP unit less the thermal energy input required to regenerate the liquid desiccant solution. The energy input to the tri-generation system is the required fuel energy input $(\dot{Q}_{\mathrm{CH_4}})$ to the SOFC CHP unit.

The results from the tri-generation system analysis are presented in three sections. First, Sect. 4.3.1 presents a specific case based on the thesis technical objectives. The specific case will provide a theoretical benchmark value for the first of its kind SOFC liquid desiccant tri-generation system. Following this, Sect. 4.3.2 provides a parametric analysis to investigate the effect changes in electrical and cooling capacities have on the overall tri-generation system performance. The aim of this is to understand the interaction between sub component operation and overall tri-generation system performance. A 24 h tri-generation simulation is also provided. Finally, Sect. 4.3.3 presents a tri-generation system climatic performance investigation. The aim of this is to evaluate the performance of the novel system under ‘real’ operating conditions to see how it performs in different geographical locations. Throughout the analysis, the SOFC tri-generation system performance is compared to a conventional separated system comprising grid electricity, gas fired boiler and vapour compression cooling system.

4.3.1 Specific Tri-generation System Case

This section presents a specific case based on defined selection criteria. As highlighted in Sect. 4.2, the SOFC CHP system selection criteria are based on the technical objectives discussed in Chap. 1 and realistic operational values. For tri-generation system optimisation it is essential that three criteria are met: (1) the technical objectives, (2) realistic operating values, and (3) a thermal agreement between the SOFC CHP system and liquid desiccant regenerator. Table 4.4 provides the selected input values for the specific novel tri-generation system. The input values used for the SOFC CHP system are the same as those used in Sect. 4.2.1.2 due to little or no variation in realistic operational values for the SOFC whilst maintaining the technical objectives of the thesis. As a result, it is the liquid desiccant system's operation that has been optimised for successful tri-generation system integration. There are more variables in a liquid desiccant system that may be controlled in a realistic range. This conclusion is demonstrated during experimental tri-generation system integration presented in Chap. 7. Unless listed in Table 4.4, the operating values used for the SOFC CHP and liquid desiccant system are those used previously and provided in Tables 4.1 and 4.3 respectively.

The water volumetric flow in the SOFC CHP WHR circuit has been selected based on achieving the highest outlet desiccant solution temperature, as shown in Fig. 4.10.

In the specific tri-generation system case a desiccant solution mass concentration of 0.65 and solution volumetric flow of 2.75 L min^{-1} are used. These values have been selected due to maximising the utilisation of thermal energy from the SOFC CHP system, and thus maximising the cooling output. As shown in Fig. 4.10, at a solution volumetric flow of 2.75 L min^{-1} and an assumed 45 °C water return temperature, the SOFC CHP WHR circuit heats the desiccant solution to a temperature of 44 °C. As demonstrated in Fig. 3.25 as desiccant solution mass concentration increases so does the required solution temperature to facilitate balanced regeneration. At a 44 °C solution temperature, the maximum operating desiccant solution mass concentration is 0.65. The performance of the specific tri-generation system case is presented in Table 4.5.

Table 4.4 Selected input values for specific tri-generation system case

Parameter	Value	Parameter	Value
SOFC current density (mA cm^{-2})	600	Air volumetric flow (m^3 h^{-1})	191
SOFC temperature (°C)	800	Des vol. flow (L min^{-1})	2.75
Number of cells in stack	53	Des mass concentration	0.65
Fuel utilisation factor (%)	80	WHR volumetric flow (L min^{-1})	2.25

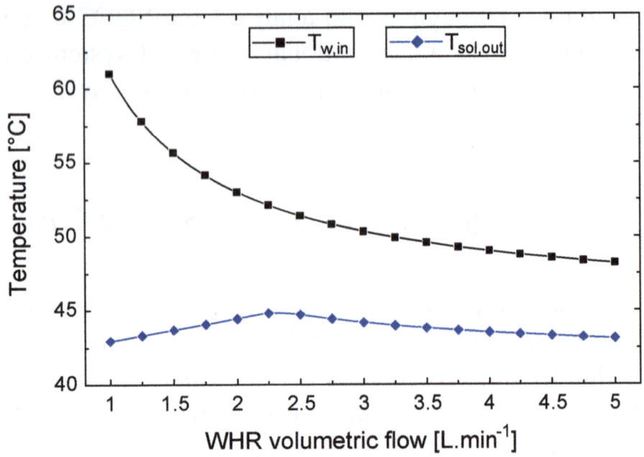

Fig. 4.10 WHR circuit water volumetric flow optimisation

Table 4.5 Performance data of specific tri-generation system case

Parameter	Value	Parameter	Value
$\dot{W}_{elec,AC}$ (W)	1526	$T_{sol,reg}$ (°C)	44
$\dot{Q}_{WHR}$ (W)	1116	$\dot{Q}_{reg}$ (W)	1108
$\dot{Q}_{CH_4}$ (W)	3345	η_{elec} (%)	45.63
$\dot{Q}_{cooling}$ (W)	920	η_{CHP} (%)	78.98
MRR (g s^{-1})	0.3168	η_{tri} (%)	70.07
$T_{a,out}$ (°C)	28.2	COP_{th}	0.83
$RH_{a,out}$ (%)	56.2	COP_{el}	8.4

The results presented in Table 4.5 demonstrate that the novel system can reach 79 % in co-generation heating (CHP) mode and 70 % in tri-generation cooling mode. Because the COP_{th} of the desiccant system is less than one and the desiccant unit has a small electrical load, CHP efficiency is higher than tri-generation. In tri-generation cooling mode, the selected (specific) system maximises the utilisation of thermal energy from the SOFC CHP system in the liquid desiccant system. A lower cooling capacity could have been selected, thus elevating tri-generation system efficiency; however the system would be more akin to a CHP operating scenario. The selected values fulfil two of the technical objectives of the thesis, namely a 1.5kW_e system operating at an electrical efficiency of 45 % or higher, and the operating values for both the SOFC CHP and liquid desiccant air conditioning system are within a realistic range. Furthermore, the system produces a meaningful amount of cooling (almost 1 kW). However, the overall system efficiency is lower than the 85 % target. This is primarily due to a liquid desiccant system COP_{th} of less than one (1.18 would be required). The inclusion of liquid desiccant air conditioning provides an efficiency increase of up to 24 % compared to SOFC electrical operation only, demonstrating the potential of the novel tri-generation system in applications that require simultaneous electrical power, heating and dehumidification/cooling.

Table 4.6 Percentage difference in PED, operating cost and emissions between the novel tri-generation system and an equivalent conventional separated system

Tri-generation system operating mode	Δ% PED	Δ% cost	Δ% emissions
CHP	−42.96	−55.25	−42.75
Tri-generation cooling	−41.25	−56.41	−63.21

Electrical import cost and emission factor = 0.172 £ kWh^{-1} (Goot 2013) and 0.555 kg CO_2 kWh^{-1} (AMEE 2014)/natural gas import cost and emission factor = 0.0421 £ kWh^{-1} and 0.184 kg CO_2 kWh^{-1} (EST 2014)

The energetic performance of the novel SOFC tri-generation system has been compared to a conventional separated system comprising: grid electricity, gas fired boiler and electrical driven vapour compression cooling, using an adapted version of the tri-generation primary energy savings energy methodology discussed in Sect. 2.6 (Badami et al. 2012). The electrical efficiency of the conventional separated system has been assumed as 33 %, a figure considering the efficiency of utility scale electrical generation plus transmission losses (Wu et al. 2014). The thermal efficiency of the gas fired boiler has been assumed as 90 %. The COP_{el} of the vapour compression cooling system is assumed constant at 2 (Welch 2008). Thus, the overall efficiency of the conventional separated system can be calculated for any given electrical, heat and cooling output from the SOFC CHP/tri-generation system.

For the outputs listed in Table 4.5 the conventional separated system has an overall efficiency of 45.05 % in CHP mode and 41.17 % in tri-generation cooling mode. The relative performance, namely Primary Energy Demand (PED), cost and emission reductions generated by the novel tri-generation system compared to the equivalent conventional separated system are presented in Table 4.6.

Compared to the conventional separated system, the novel tri-generation system demonstrates a percentage reduction in PED of 43–45 %, an operating cost reduction of 55–59 %, and an emission reduction of 43–65 %. The encouraging PED, cost and emission reductions are primarily due to the high electrical efficiency of the SOFC and the replacement of electrically derived cooling with waste heat driven cooling. Because grid electricity has a high economic cost and emission factor, generating it at high electrical efficiencies from lower cost and emission factor natural gas produces great benefit compared to a conventional separated system. A comprehensive economic and environmental assessment is provided in Chap. 8.

Wu et al. (2014) theoretically and experimentally evaluated a micro tri-generation system incorporating a 16kWe ICE with a vapour absorption chiller. Experimental results demonstrate a tri-generation system efficiency of 76.5 % in CHP mode and 56 % in cooling mode. Compared to a conventional separated system, a clear efficiency increase was gained in CHP mode, however only a marginal improvement was observed during cooling mode. This is because of a low COP_{th} of the vapour absorption chiller (~0.4). Compared to other theoretical assessments of micro tri-generation systems employing either SOFC or liquid desiccant technology, the theoretical analysis of the novel SOFC liquid desiccant tri-generation system shows promising results. This is primarily due to the high electrical efficiency of the SOFC (~45 %) and reasonable COP_{th} of the liquid desiccant air conditioning system (~0.8).

Next, Sect. 4.3.2 presents a tri-generation system parametric performance analysis.

4.3.2 *Tri-generation System Parametric Analysis*

In this section, a parametric analysis is presented; investigating the effect changes in the electrical and cooling capacity of the tri-generation system have on overall efficiency and PED reduction. The aim of this is to understand the interaction between sub component capacity and overall tri-generation system performance. A 24 h tri-generation system simulation is also provided. Unless otherwise varied, the baseline input values for the SOFC CHP and liquid desiccant system are those provided in Tables 4.1 and 4.3 respectively.

Figure 4.11 demonstrates the impact SOFC electrical capacity has on tri-generation system performance. Desiccant system capacity is kept constant throughout. The parametric analysis assesses tri-generation performance at a 0.65 and 0.7 desiccant solution mass concentration. As discussed in Sect. 4.3.1, when operating with the assumed 45 °C return water temperature in the SOFC WHR circuit, a 0.65 solution mass concentration is optimal to facilitate balanced regeneration. For the system to operate with a 0.7 solution mass concentration the required heated solution temperature is 48 °C, and thus requires a WHR return water temperature of around 51 °C. This return water temperature has been assumed attainable for comparative purposes only during the parametric analysis.

Figure 4.11a demonstrates the impact SOFC current density has on tri-generation system efficiency and PED reduction. As current density is increased from

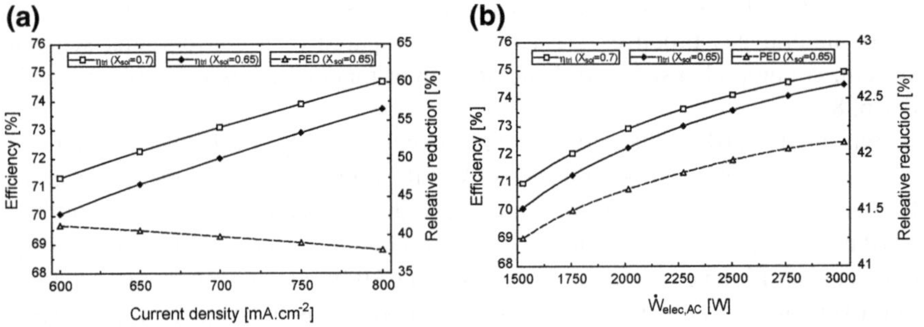

Fig. 4.11 Variation in SOFC tri-generation system performance with **a** SOFC current density, and **b** SOFC electrical power output

600 to 800 mA cm^{-2} tri-generation system efficiency increases from 70.07 to 73.74 % at a desiccant solution mass concentration of 0.65 and 71.33–74.69 % at a 0.7 desiccant solution mass concentration. As current density is increased, with cell number and cell size held constant, the electrical efficiency decreases and the available thermal output increases. As a result, a smaller proportion of the thermal energy output is used for desiccant solution regeneration, and because the desiccant system has a COP_{th} of 0.83–0.86 i.e. less than one, tri-generation efficiency increases. A marginal increase in tri-generation efficiency is observed at a higher desiccant solution mass concentration; this is due to a greater cooling output for the same given thermal input. As the current density increases at a desiccant solution mass concentration of 0.65, the percentage reduction in PED compared to the equivalent conventional separated system reduces from 41.25 to 38.03 %. This is because the electrical efficiency of the SOFC reduces with current density, and thus the electrical efficiency advantage (electricity has a higher associated PED) over the equivalent conventional separated system is diminished.

Figure 4.11b demonstrates the impact increasing the SOFC electrical capacity from 1500 We to 3000 We has on tri-generation system efficiency and PED reduction. SOFC electrical capacity is increased by increasing the number of SOFC cells from 53 to 105. As the electrical capacity of the system increases so does tri-generation system efficiency from 70.7 to 74.49 % at a desiccant solution mass concentration of 0.65 and 70.97–74.94 % at a desiccant solution mass concentration of 0.7. Similar to the current density investigation, as the electrical capacity of the tri-generation system is increased, more thermal energy is produced, and a smaller proportion is used for desiccant solution regeneration, resulting in higher tri-generation efficiency. As the electrical capacity of the system increases at a desiccant solution mass concentration of 0.65, the percentage reduction in PED compared to the equivalent conventional separated system increases from 41.25 to 42.11 %.

Figure 4.12 demonstrates the impact dehumidifier cooling capacity, at a desiccant solution mass concentration of 0.65, has on tri-generation system efficiency and PED reduction. Cooling capacity is increased by increasing the desiccant solution volumetric flow from 0.31 to 2.8 L min^{-1}. The electrical and thermal output of the SOFC has been kept constant throughout. Tri-generation system efficiency increases from 72.29 % at a 500 W cooling capacity to a peak efficiency of 73.58 % at a 700 W cooling capacity. Tri-generation system efficiency then decreases to 69.96 % at the maximum cooling output achievable of 924 W. There is a marginal change in the percentage reduction in PED compared to the equivalent conventional separated system. The PED percentage reduction follows a similar trend to the tri-generation system efficiency, but with a peak reduction of 41.77 % at an 800 W cooling capacity.

Using Nottingham summer climatic data (temperature and relative humidity) a dynamic simulation of the novel tri-generation system operating over one daily summer cycle is presented in Fig. 4.13. The operating values used for the tri-generation system simulation are those provided in Table 4.4. The assessment assumes an air conditioning load between 08:00 and 20:00. Outside of the air

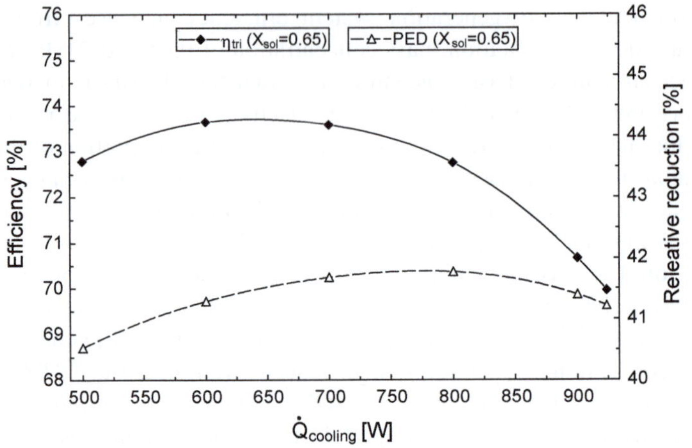

Fig. 4.12 Variation in SOFC tri-generation system performance with dehumidifier cooling output

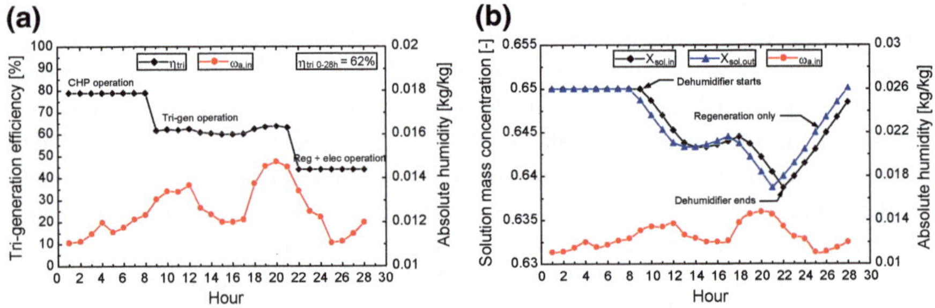

Fig. 4.13 24 h simulation of the novel tri-generation system operating in a UK climate

conditioning load requirement the tri-generation system operates in CHP mode. An initial desiccant solution mass concentration of 0.65 has been assumed. A 0.65 desiccant solution mass concentration prerequisite provides an average supply air condition of 21 °C and 60 % RH over the simulation period, an acceptable value for a UK summer air conditioning application. The aim of the simulation is to finish the daily cycle with the same starting desiccant solution mass concentration so that the process may be repeated.

Figure 4.13a shows that between 00:00 and 08:00 the tri-generation system is operating in CHP mode, with a system efficiency of 79 %. Figure 4.13b shows that during this period the desiccant solution mass concentration remains constant. This is because the dehumidifier/regenerator is not operating. At 08:00 the tri-generation system is switched to cooling mode (dehumidifier and regenerator). Figure 4.13a shows that during this period tri-generation efficiency ranges between 60 and 64 %. Figure 4.13b shows that during the air conditioning period the desiccant solution mass concentration falls. This is because the thermal input from the SOFC is not sufficient to produce an instantaneous regenerator MAR that balances the dehumidifier MRR. As a result by 20:00 the solution mass concentration has been diluted to 0.6387. In order to re-store the 0.65 solution mass concentration the tri-generation system needs to operate in regeneration (plus electrical output) mode from 20:00 onwards. Figure 4.13b shows that it takes until hour 28 (04:00 the next day) for the desiccant solution to be re-stored to a 0.65 mass concentration. Figure 4.13a shows that the tri-generation system efficiency during this period is 44 %. The overall efficiency of the tri-generation system over the whole 28 h time period is 62 %. The tri-generation simulation has established that the novel system can operate in a balanced manner over a daily cycle (less than 24 h from the start of the air conditioning load to restoration of the solution mass concentration), with reasonable system performance demonstrated.

Section 4.3.2 has presented a tri-generation system parametric analysis, investigating the impact changes in electrical and cooling capacities have on the overall system performance. Across all values investigated, the novel tri-generation system demonstrates promising performance figures. It is evident that tri-generation system performance is highest when a smaller proportion of the SOFC's thermal output is used for desiccant solution regeneration. However, to generate a meaningful amount of cooling this should not necessarily dictate system operation. Next, Sect. 4.3.3 presents a tri-generation system climatic performance analysis.

4.3.3 Tri-generation System Climatic Performance Investigation

In this section, the performance of the novel SOFC tri-generation system is assessed for three different (climatic) locations: Nottingham, Porto and Hong Kong. These three locations have been selected due to their diverse, notably temperature and relative humidity, climatic conditions, therefore providing a comprehensive evaluation of the novel tri-generation system in a global context. The summer design condition, dry bulb temperature and relative humidity, have been acquired from the EnergyPlus weather tool. The input values for the SOFC CHP and liquid desiccant system are those used in the specific case presented in Sect. 4.3.1. A desiccant solution mass concentration of 0.65 has been used in all scenarios.

Table 4.7 presents the results from the climatic performance investigation. The climatic assessment has aimed to maximise the use of the SOFC thermal output for desiccant solution regeneration and to demonstrate the maximum possible cooling output/tri-generation efficiency in each location at the 0.65 solution mass concentration. The operational desiccant solution mass concentration has remained constant between locations. As a result, the supply air condition for each geographical case is not the same. Because the tri-generation system operating in CHP mode is not influenced by the summer design condition, the CHP

Table 4.7 Tri-generation system climatic performance investigation results

	Nottingham (UK)	Porto (PT)	Hong Kong (CN)
$T_{a,in}$ (°C)	26.7	30.3	33.8
$RH_{a,in}$ (%)	42.1	35.2	56.7
$T_{a,out}$ (°C)	23.28	25.95	30.9
$RH_{a,out}$ (%)	41.45	37.74	49.24
$\dot{Q}_{cooling}$ (W)	507	524	965
η_{tri} (%)	57.52	58.03	71.3
Δ% PED	33.9	34.23	41.9
Δ% Cost	50.97	51.21	56.9
Δ% Emissions	60.47	60.59	63.46

performance is equal to that presented in Tables 4.5 and 4.6, and only tri-generation performance data is presented. The cost and emission factors used are based on those listed in Table 4.6.

It is evident from the results presented in Table 4.7, that tri-generation system efficiency, PED, cost and emission percentage reductions are highest in the Hong Kong location and lowest in Nottingham. For the given desiccant system operating conditions the cooling output is larger, and thus the tri-generation system performance is greater. Elevated cooling output in Hong Kong is due to a more humid climate and thus a greater mass transfer potential at the given 0.65 solution mass concentration.

Section 4.3 has presented a comprehensive theoretical analysis of the proposed novel SOFC tri-generation system. A specific case based on defined selection criteria, parametric analysis and climatic performance investigation have all been presented. The results demonstrate that the integration of SOFC and liquid desiccant into a complete tri-generation system for building applications is feasible, and shows great potential with high tri-generation system efficiency attainable at low system capacities. Next, Sect. 4.4 presents the chapter conclusions and their implications for the experimental system.

4.4 Chapter Conclusions and Implications for Experimental Systems

This chapter has presented a theoretical assessment of the first of its kind SOFC liquid desiccant tri-generation system, operating with a potassium formate desiccant solution. The aim of this chapter has been to demonstrate the potential for the development of an efficient and effective tri-generation system based on SOFC and liquid desiccant air conditioning technology. Objectives of the theoretical analysis have been to quantify and qualify the benefits of combining the two technologies, to provide a comparison of performance compared to a conventional separated system and to offer a benchmark of performance/operational considerations for the experimental tri-generation system.

Using validated models, Sect. 4.2 has analysed the SOFC CHP and liquid desiccant air conditioning sub components. The aim has been to improve the understanding of the sub-component operation and select specific sub-component operating values to facilitate effective tri-generation system integration. Conclusions from the sub-component investigation include:

- Whilst remaining within the boundaries of realistic operating values, it has been demonstrated that there are more operational variables that may be controlled in the liquid desiccant air conditioning component compared to the SOFC CHP component. As a result, it is the desiccant system's operating values that need to be optimised for successful tri-generation system integration. This is a conclusion relevant to experimental tri-generation system integration presented in Chap. 7.
- Potassium formate is an appropriate choice of working fluid for the desiccant air conditioning system due to its low temperature regeneration requirement, which demonstrates good thermal agreement with the SOFC CHP systems outlet water temperature.
- The liquid desiccant air conditioning system demonstrates effective use of low grade thermal energy (<60 °C). COP_{th} values of 0.8 are attainable at a regenerator solution temperature of 40–50 °C.
- As the working solution mass concentration of the potassium formate solution increases so does the requirement for a higher solution temperature in the regenerator, and thus a higher outlet water temperature from the SOFC CHP system. Matching of these two variables is essential for successful tri-generation system integration. At the assumed 45 °C return water temperature in the SOFC WHR circuit, a potassium formate solution mass concentration of 0.65 is appropriate to facilitate balanced desiccant system operation. Increasing the potassium formate solution mass concentration to 0.7 requires a return water temperature of 51 °C, but only provides a marginal increase in the tri-generation system efficiency of around ~1.26 % (max).

Section 4.3 has combined the SOFC CHP and liquid desiccant sub-component models into a complete tri-generation system model. A specific case based on defined selection criteria, parametric analysis and climatic performance investigation have all been presented. Conclusions from the tri-generation system modelling and analysis include:

- The pairing of SOFC and liquid desiccant air conditioning technology is a viable option, and shows great potential for the development of an efficient and effective tri-generation system. Due to the high electrical efficiency of the SOFC and reasonable COP_{th} of the liquid desiccant air conditioning system (at the proposed system capacities), the novel tri-generation system demonstrates high overall efficiency even in cooling mode.

- Based on the thesis technical objectives, and the use of realistic operating values, simulations of the novel tri-generation system demonstrate an overall efficiency of 78.98 % in CHP mode and 70.07 % in tri-generation cooling mode. Values competitive with other micro tri-generation systems presented in the literature. Furthermore, encouraging PED, cost and emission reductions compared to a conventional separated system have been demonstrated.
- The inclusion of liquid desiccant air conditioning provides an efficiency increase of up to 24 % compared to SOFC electrical operation only, demonstrating the merit of the novel tri-generation system in applications that require simultaneous electrical power, heating and dehumidification/cooling.
- Tri-generation performance is largely influenced by the sub-components system capacity. Tri-generation efficiency increases as the proportion of the SOFC thermal output used for desiccant solution regeneration is reduced; this is due to a COP_{th} of less than one.
- The novel tri-generation system performs well across a range of climatic conditions. For the given conditions and evaluation method tri-generation efficiency and PED reduction are highest in hotter more humid climates such as Hong Kong.

This chapter has provided a clear contribution to knowledge through the theoretical analysis of the first of its kind SOFC liquid desiccant tri-generation system. The work presented has demonstrated that the technological paring is viable, and has the potential to deliver higher overall efficiency.

References

Al-Sulaiman, F.A., I. Dincer, and F. Hamdullahpur. 2010. Energy analysis of a trigeneration plant based on solid oxide fuel cell and organic Rankine cycle. *International Journal of Hydrogen Energy* 35(10): 5104–5113.

AMEE. 2014. Realtime carbon. Retrieved 06/04/2014, 2014, from http://realtimecarbon.org/.

Badami, M., A. Portoraro, and G. Ruscica. 2012. Analysis of trigeneration plants: Engine with liquid desiccant cooling and micro gas turbine with absorption chiller. *International Journal of Energy Research* 36(5): 579–589.

Blomen, L. J., M.J. Blomen, M.N. Mugerwa. 1993. *Fuel Cell Systems*. New York.

Blum, L., Robert Deja, Roland Peters, and Detlef Stolten. 2011. Comparison of efficiencies of low, mean and high temperature fuel cell Systems. *International Journal of Hydrogen Energy* 36(17): 11056–11067.

CIBSE. 2006. *CIBSE guide A: Environmental design*. London: CIBSE.

Elmer, T., M. Worall, S. Wu, and S.B. Riffat. 2015b. Fuel cell technology for domestic built environment applications: State of-the-art review. *Renewable and Sustainable Energy Reviews* 42: 913–931.

EST. 2014. Energy saving trust—Our calculations. http://www.energysavingtrust.org.uk/Energy-Saving-Trust/Our-calculations.

Fan, Liangdong, Chengyang Wang, Mingming Chen, and Bin Zhu. 2013. Recent development of ceria-based (nano) composite materials for low temperature ceramic fuel cells and electrolyte-free fuel cells. *Journal of Power Sources* 234: 154–174.

Goot, G. 2013. Electricity price per kWh (2013). Retrieved 06/03/2014, 2014, from http://blog.comparemysolar.co.uk/electricity-price-per-kwh-2013-comparison-of-e-on-edf-npower-british-gas-scottish-and-sse/.

Jradi, M., and S. Riffat. 2014d. Tri-generation systems: Energy policies, prime movers, cooling technologies, configurations and operation strategies. *Renewable and Sustainable Energy Reviews* 32: 396–415.

Laosiripojana, N., and S. Assabumrungrat. 2007. Catalytic steam reforming of methane, methanol, and ethanol over Ni/YSZ: The possible use of these fuels in internal reforming SOFC. *Journal of Power Sources* 163(2): 943–951.

Larmine, J., and A. Dicks. 2003. *Fuel cell systems explained*. Chichester: Wiley.

Liu, J., and S.A. Barnett. 2003. Operation of anode-supported solid oxide fuel cells on methane and natural gas. *Solid State Ionics* 158(1–2): 11–16.

Peters, R., R. Dahl, U. Klüttgen, C. Palm, and D. Stolten. 2002. Internal reforming of methane in solid oxide fuel cell systems. *Journal of Power Sources* 106(1–2): 238–244.

Pilatowsky, I., R.J. Romero, C.A. Isaza, S.A. Gamboa, P.J. Sebastian, and W. Rivera. 2011. *Cogeneration fuel cell—Sorption air conditioning systems*. London: Springer.

Tuyen, N.D., and G. Fujita. 2012. Modelling a SOFC power unit using natural gas fed directly. In *Advances in natural gas technology*, ed. H.A. Al-Megren. Rijeka: INTECH.

Welch, T. 2008. In *CIBSE knowledge series: KS13—Refrigeration, ed.* H. Carwarardine and K. Butcher. CIBSE Publications. ISBN: 978-1-903287-91-0. London

Wu, J.Y., J.L. Wang, S. Li, and R.Z. Wang. 2014. Experimental and simulative investigation of a micro-CCHP (micro combined cooling, heating and power) system with thermal management controller. *Energy* 68: 444–453.

Chapter 5
Component Evaluation of a Novel Integrated Liquid Desiccant Air Conditioning System (IDCS)

5.1 Introduction

This chapter presents component evaluation, based on experimental data, of a novel 'integrated' liquid desiccant air conditioning system (IDCS) operating with a potassium formate solution. It evaluates the novel system under a variety of environmental and operating conditions in a controlled laboratory environment, with the aim of facilitating effective tri-generation system integration in Chap. 7.

As previously discussed in Chap. 2, there are some clear operational advantages to the use of liquid desiccant air conditioning systems over conventional VCS in building applications. The advantages are a low electrical requirement, the replacement of electrical energy with thermal energy (this can be waste or process heat) and high latent control. Liquid desiccant air conditioning also offers significant advantages over other thermally activated air conditioning technologies, particularly in a tri-generation system application. The advantages are a potential to use low grade waste heat at temperatures of less than 60 °C and operation at atmospheric pressure. Although extensive work has been carried out on liquid desiccant air conditioning systems (Halliday et al. 2002; Gommed and Grossman 2007; Beccali et al. 2012; Hassan and Mohamad 2012), to date system complexity has severely limited their wider application and outweighed the significant energy savings they can achieve (Jain and Bansal 2007). Furthermore, space, complexity and leakage are often cited as a significant barrier to the wider use of liquid desiccant air conditioning in building and tri-generation system applications (Jain and Bansal 2007). There is therefore a need for simpler, more compact systems, particularly for building and tri-generation system applications. Additionally, across all applications, issues of liquid desiccant entrainment in the supply airstream and the risk of HVAC equipment corrosion need to be addressed.

T. Elmer, *A Novel SOFC Tri-generation System for Building Applications*,
Springer Theses, DOI 10.1007/978-3-319-46966-9_5

A novel IDCS has been developed with the aim of addressing the highlighted issues, therefore permitting effective integration of liquid desiccant air conditioning technology in building and tri-generation system applications. The novel IDCS presented in this chapter has three design characteristics that aim to address the issues of size, leakage and corrosion:

(1) A novel stack design integrates the regenerator, evaporative inter-cooler and dehumidifier into a single heat and mass exchanger (HMX) core, making the whole system more compact and less prone to leakage. The IDCS has less piping, heat exchangers and pumps compared to an equivalent conventional 'separate' system.
(2) The use of a semi-permeable micro porous membrane in the dehumidifier and regenerator HMX cores to prevent desiccant entrainment into the supply airstream.
(3) Employment of an environmentally friendly, non-corrosive and low cost potassium formate ($CHKO_2$) desiccant solution.

No previous work has been found in the literature regarding an integrated design combining the regenerator, evaporative inter-cooler and dehumidifier into a single HMX core, thus the work presented is a clear contribution to knowledge. The work presented in this chapter progresses the fields of liquid desiccant air conditioning technology for building and tri-generation system applications. The IDCS has been acquired from the company Intelligent Integration of Solar Air and Water (IISAW) in China, but has been tested at The University of Nottingham. Detailed component testing of the IDCS under controlled laboratory conditions generates greater understanding of the system and its performance, but more specifically it facilitates the following:

(1) Highlight the impact changes in environmental and operating conditions have on the performance of the dehumidifier, regenerator and complete IDCS.
(2) Recording of the operating conditions in which the dehumidifier and regenerator processes are balanced i.e. the mass of water absorbed in the dehumidifier is equal to the mass removed in the regenerator. This will allow continuous operation of the IDCS across a range of working requirements.
(3) Discuss the potential for tri-generation system integration. Selection of optimised operating conditions for the IDCS, with particular consideration to tri-generation system integration i.e. hot water flow rate and temperature from the SOFC CHP system.

The chapter is presented in three sections. Section 5.2 describes the IDCS experimental set-up, instrumentation and experimental method. Section 5.3 presents the IDCS experimental results and analysis. Section 5.4 concludes the chapter and discusses the potential for tri-generation system integration. A published journal article, by the author, on the novel integrated liquid desiccant air conditioning system can be referred to here; Elmer et al. (2016c).

5.2 IDCS Experimental Set-up

This section provides a detailed description of the experimental set-up, instrumentation and experimental method used in the evaluation of the IDCS.

As previously stated, space, complexity and leakage is often cited as a significant barrier to the wider use of liquid desiccant air conditioning in buildings. As a result, an efficient and compact liquid desiccant system has been designed and built. The regenerator (R/C), evaporative inter-cooler (E/C) and dehumidifier (D/C) cores are all combined into one single HMX core. The membrane based HMX runs the entire length of the unit, but is subdivided into three different airflows, and two different fluid flows; desiccant and water. Thermal input to the regenerator is achieved through the heating of the inlet airstream in a liquid to air heat exchanger. Figure 5.1 shows a schematic of the integrated unit design concept. This design significantly reduces the number of heat exchangers, pipes and ducting often seen in liquid desiccant air conditioning systems, therefore reducing its total footprint.

The novel stack-integrated design has three distinct advantages:

1. More compact form, essential for buildings applications.
2. Reduced risk of desiccant leakage.
3. Prevention of desiccant carry-over into supply air stream.

Figure 5.2 provides a labelled schematic diagram of the complete IDCS connected to a hot water cylinder.

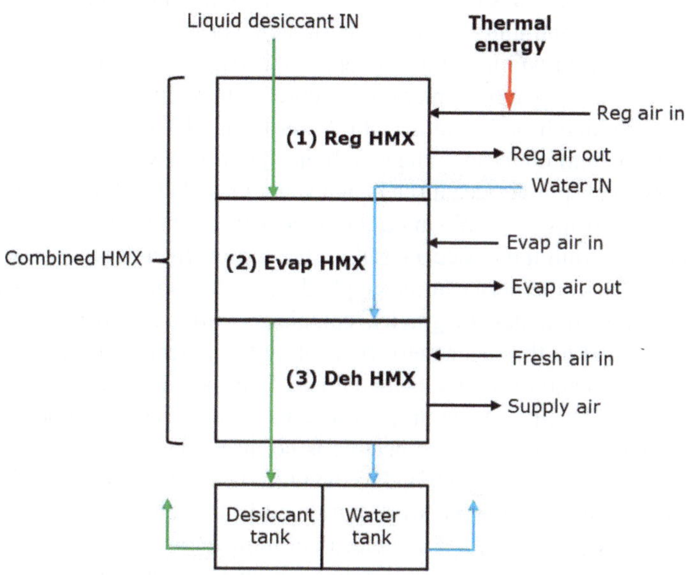

Fig. 5.1 The novel IDCS concept

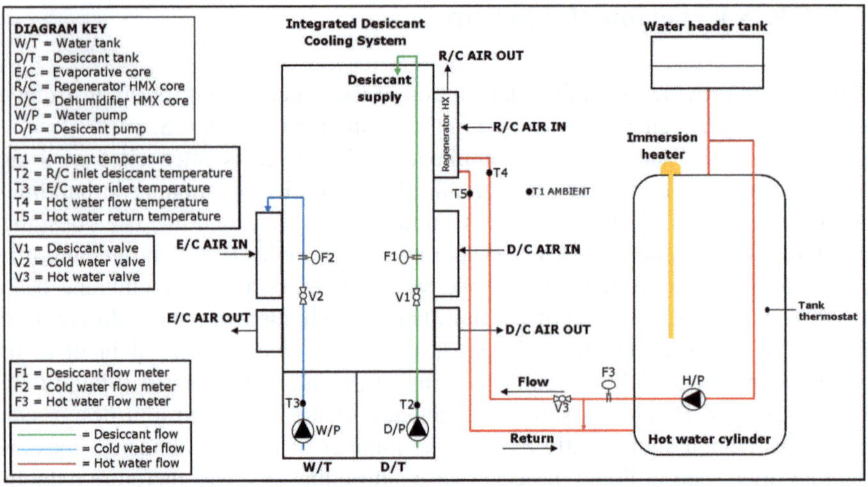

Fig. 5.2 System schematic of the laboratory testing of the IDCS

The IDCS HMX core consist of 26 channels that allow air and desiccant solution/water to flow in a cross flow manner (air through the core, desiccant/water downwards through the core), separated by a semi permeable micro-porous membrane. The solution channels consist of a polyethylene sheet, with membranes attached on either side. The gap between the two solution channels provides the space for the air to flow. The membrane allows the diffusion of water vapour, but prevents liquid desiccant solution migration across it, thus overcoming the issue of liquid desiccant entrainment in the air stream. The regenerator core is 310 mm in height, 420 mm in width and has a depth of 240 mm, with 26 air channels. The dehumidifier and evaporative inter-cooler core is 695 mm in height, 420 mm in width and has a depth of 240 mm, with 26 air channels. The entire HMX core is contained in an aluminium box. The membrane based HMX core sits on top of a 20 litre stainless steel split desiccant (D/T) and water tank (W/T). Weak desiccant solution is pumped, using a 15 W single phase centrifugal magnetically driven pump (0–10 L min^{-1}), from the desiccant tank to the top of the unit where the regenerator is located. Here the desiccant is supplied through a spray nozzle, and flows in a downward direction due to gravity, contained within the membrane. Thermal energy is supplied to the regenerator by heating the regenerator airstream prior to it entering the regenerator HMX using a liquid to air heat exchanger. Heating of the airstream lowers the air side vapour pressure, and thus drives mass transfer from the desiccant solution. Direct solution heating is not used due to the integrated design. The regenerator airstream is supplied to the unit via a 500 m^3 h^{-1} (nominal) 240 V AC axial fan. The experimental work presented in this chapter uses a vented 120 litre hot water cylinder with a 3 kW electrical immersion heater as the regenerator heat source. However, the electrical immersion heater could be replaced with any heat source that can provide hot water at the desired temperature and flow rate i.e.

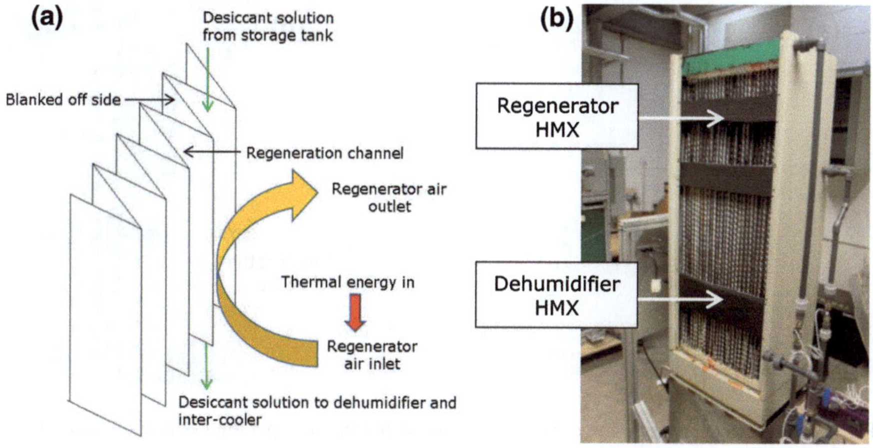

Fig. 5.3 **a** Regenerator core operating concept, and **b** the regenerator HMX

SOFC. A Wilo-Smart A-rated 230 V AC pump has been employed to circulate the hot water in the heating circuit. A Honeywell L641A cylinder thermostat has been used to maintain the flow temperature from the tank at a constant temperature. The heated regenerator air stream then passes across the desiccant soaked membrane causing the dilute desiccant solution to be re-concentrated due to the removal of water by vaporisation into the regenerator air stream. The liquid desiccant leaves the regenerator as concentrated (strong) solution. The structure of the regenerator core is shown in Fig. 5.3a and a photograph in Fig. 5.3b. One side of the regenerator exchanger is blanked off. This is because in the regeneration process only one airstream is required, that to regenerate the desiccant solution. However, in the lower section of the IDCS there are two air processes, evaporative cooling and dehumidification, and so two air channels are required, as shown in Fig. 5.4a.

After the regeneration process, the desiccant solution flows downwards due to gravity through the desiccant evaporative inter-cooler and dehumidifier. Here two processes occur simultaneously (1) an evaporative cooling process creates a sensible cooling effect, which is transferred across the HMX wall to cool the desiccant and supply air stream, and (2) the supply air stream is dehumidified and cooled. The structure of the desiccant evaporative inter-cooler and dehumidifier core is shown in Fig. 5.4a and a photograph in Figs. 5.3b and 5.4b. The evaporative cooling process is not only advantageous for the lowering of the supply air temperature; it also removes the latent heat of condensation produced during the dehumidification process and creates a lower vapour pressure in the desiccant solution, and thus a greater dehumidification potential.

Water is pumped using a 15 W single phase centrifugal magnetically driven pump (0–10 L min^{-1}) from the water tank to the top of the evaporative core and is supplied through a spray nozzle. On the evaporative side of the HMX, water flows downwards due to gravity over the exchanger surface. The evaporative cooler airstream

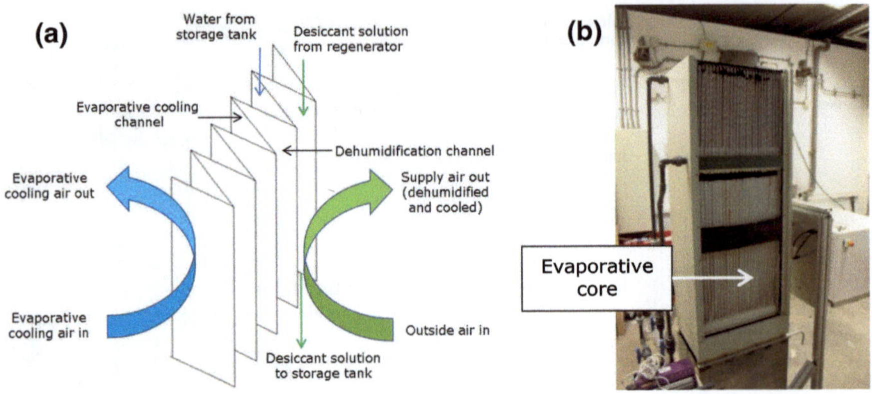

Fig. 5.4 **a** Desiccant evaporative inter-cooler and dehumidifier operating concept, and **b** photograph of dehumidifier

is supplied via a 500 $m^3\ h^{-1}$ (nominal) 240 V AC axial fan. This air flows across the HMX in a cross-flow manner. This causes direct evaporative cooling and indirectly cools, through the exchanger wall, the liquid desiccant solution and dehumidifier supply air stream. Because the evaporative cooling and dehumidification processes are separated by the exchanger wall, sensible cooling is provided to the desiccant solution and supply airstream without moisture addition. On the dehumidifier side of the HMX, fresh air is supplied to the HMX core via a 500 $m^3\ h^{-1}$ (nominal) 240 V AC axial fan. The fresh air flows in a cross flow manner across the desiccant soaked membrane. Due to the lower temperature (evaporatively cooled) and vapour pressure (regeneration) of the desiccant solution, the air is sensibly cooled and dehumidified. This air can then be supplied directly to the room, or can be passed through another evaporative cooling process to lower its temperature further. The warm and weak desiccant solution then flows back to the desiccant tank to begin the process again. The water used in the evaporative inter-cooler flows back to the separate tank.

Many of the liquid desiccant systems reported in the literature heat and cool the desiccant solution directly prior to the regeneration and dehumidification process respectively. However, in this IDCS because all desiccant flow is contained within one complete HMX, the desiccant solution cannot be extracted for prior heating/ cooling, thus heating of the regeneration air stream and the inclusion of an evaporative inter-cooler is required.

Next, Sect. 5.2.1 provides details of the instrumentation used in the experimental testing of the IDCS.

5.2.1 Instrumentation

This section will describe the instrumentation used on the IDCS. Reference to the IDCS schematic in Fig. 5.2 is made throughout. All fans on the IDCS are equipped with Vent Axia infinitely variable fan speed controllers to enable control of the volumetric air flow through the cores. The air inlet and outlets of the regenerator, evaporative and dehumidifier cores are fitted with 125 mm galvanised steel spiral tube ducting. The inlet and outlet air flows are instrumented with Vaisalia HMP110 humidity and temperature probes. The probes are mounted within the spiral tube ducting using special flanges. The humidity and temperature probes are factory calibrated. Air velocity through the dehumidifier, regenerator and evaporative inter-cooler cores are measured using an RS AM4204 hot wire anemometer at the air ducting outlets. The hot wire anemometer is factory calibrated. Air velocity measurements are recorded at five points across the air duct, and the average taken. The air velocity measurements are also validated against a TSI LCA501 rotating vane anemometer.

The desiccant and water pipes connecting the tank to the core have been equipped with ball valves (V1 and V2) so that the desiccant or water volumetric flow rate may be set to a desired value. A valve has also been placed on the hot water circuit (V3) to control the hot water flow to the regenerator. All water and desiccant solution flows have been instrumented with sheathed K-Type thermocouples (Nickel Chromium/Nickel Aluminium). Thermocouples have been placed at the inlet to the desiccant side (T2) and water side (T3) of the HMX core. Thermocouples have also been placed at the hot water inlet (T4) and outlet (T5) to the regenerator liquid to air heat exchanger.

The desiccant solution and water volumetric flow is measured using a 0.2–2 L min^{-1} Parker Liquid Flow Indicator. These are placed on the pipe connecting the tank to the HMX core (F1 on desiccant side, and F2 on water side). The flow meters used are calibrated for water at 20 °C according to density and viscosity. Thus, for the water flows used in the system, no correction is required. For the desiccant solution flow a correction factor is required to equate the volumetric flow shown on the flow meter, to the actual desiccant flow. This correction correlation is shown in Eq. 5.1 (Liu 2008).

$$v_{sol} = v_{\text{w}} \sqrt{\frac{\left(m_{\text{float}} - V'_{\text{float}} \rho_{\text{des}}\right) \rho_{\text{w}}}{\left(m_{\text{float}} - V'_{\text{float}} \rho_{\text{w}}\right) \rho_{\text{des}}}} \tag{5.1}$$

v_{sol} and v_{w} is the volumetric flow in L min^{-1} of the desiccant solution and water respectively. For the flow meters used, the float weight, $m_{\text{float}} = 2.1 \times 10^{-3}$ kg and the float volume $V'_{\text{float}} = 0.25 \times 10^{-6}\ \text{m}^3$. Photographs of the humidity and temperature probe, K-Type thermocouple and hot wire anemometer are provided in Fig. 5.5.

The hot water cylinder is equipped with an RS 1–15 L min^{-1} piston flow meter (F3), designed for flow temperatures of up to 60 °C. All desiccant solution and water flows on the IDCS are equipped with 20 mm PVC-U plastic pipe, with plastic fittings. The hot water cylinder is piped with insulated 22 mm copper pipe and copper fittings. Flexible PVC hot water hose is used to connect the hot water cylinder to the regenerator liquid to air heat exchanger.

For the accurate evaluation of the desiccant system, the working concentration of the desiccant solution needs to be determined. Using a correlation based on

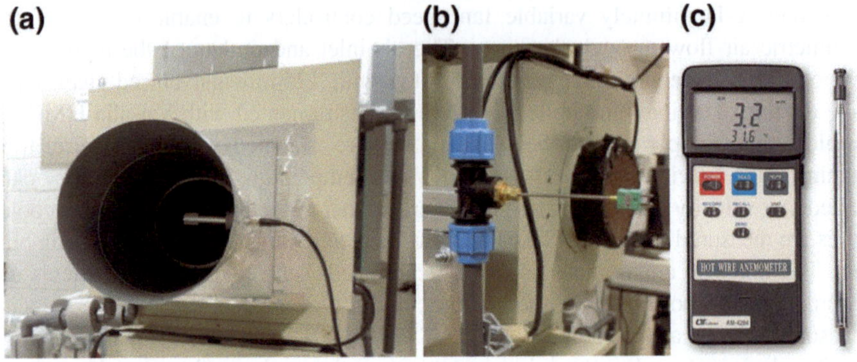

Fig. 5.5 Photographs of **a** duct mounted Vaisalia HMP110 humidity and temperature probe, **b** K-Type thermocouple, and **c**, RS hot wire anemometer

the work of Melinder (2007). The desiccant solution concentration is determined from the solution density and temperature. In the experimental work the density of the desiccant solution is measured using a differential pressure density meter with temperature compensation. The meter has been designed to work in the density range of the potassium formate solution (1400–1550 kg m^{-3}), and has been calibrated with water. The measurement prongs of the differential pressure density meter are placed in the desiccant solution tank and held until a steady-state reading is achieved. The temperature of the solution is measured using the K-Type thermocouple at the tank outlet. The concentration is then calculated using the correlation presented in Eq. 5.2.

$$\begin{aligned} X_{sol} = & -253.147703 + 0.0443853996\, T_{sol} + 0.000163666247\, T_{sol}^2 \\ & + 0.331709855\, \rho_{sol} - 0.0000793702671\, \rho_{sol}^2 \end{aligned} \tag{5.2}$$

The electrical consumption of fans and pumps are measured using a Brennenstuhl PM230 electricity monitor. This is essential for the COP_{el} calculations. At full load the desiccant system parasitic electrical load is measured at 400 W. Photographs of the liquid flow indicator, density meter and electrical monitor are provided in Fig. 5.6.

A DataTaker DT500 datalogger is used to record the data from the sensors every ten seconds. Further details of the measuring equipment used and their associated accuracy are listed in Table 5.1.

Section 5.2.1 has described the IDCS instrumentation. Next, Sect. 5.2.2 discusses the use of uncertainly analysis in the evaluation of the IDCS.

5.2.2 *Uncertainty Analysis*

Uncertainty analysis provides a measure of the error associated with a calculated value. The devices listed in Table 5.1 directly measure physical parameters against

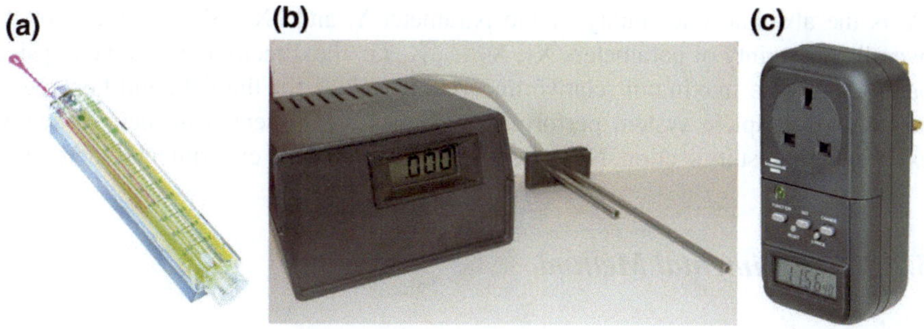

Fig. 5.6 Photographs of **a** 0.2–2 L min^{-1} Parker liquid flow indicator, **b** desiccant solution density meter, and **c** Brennenstuhl PM230 electricity monitor

Table 5.1 Instrumentation equipment and associated accuracy

Measurement device	Measurement subject	Measurement range	Measurement accuracy
HMP110 relative humidity and temperature probe	Air relative humidity and temperature	0–90 % RH 0–40 °C	$RH_a = \pm 1.7$ % RH $T_a = \pm 0.2$ °C
RS AM4204 hot wire anemometer	Air velocity	0–20 m s^{-1}	$u_a = \pm 5$ % of reading
K-Type thermocouple probe	Desiccant solution and water temperature	0–1100 °C	$T_{sol}/T_{water} = \pm 2.2$ °C
Parker Liquid Flow Indicator	Desiccant solution and water volumetric flow	0.2–2 L min^{-1}	$v_{sol}/v_{water} = \pm 2$ % of reading
Desiccant solution density meter	Desiccant solution density	1400–1550 kg m^{-3}	$\rho_{sol} = \pm 10$ kg m^{-3}
Brennenstuhl PM230 electricity monitor	IDCS electrical power usage	Up to 16 Amps	$W_{IDCS} = \pm 3$ % of reading

a suitable reference point e.g. temperature, relative humidity, air velocity, liquid volumetric flow, density and electrical power. The direct measurements taken have an associated accuracy, which may be used directly when presenting data in this form. However, other physical parameters such as absolute humidity, enthalpy and moisture removal rate cannot be measured directly, and are a function of measured physical parameters and known equations. Using the propagation of error formula shown in Eq. 5.3 (Taylor 1997), the absolute uncertainty of a calculated value can be calculated.

$$Y = f(X_1, X_2, \ldots, X_n)$$

$$\delta Y = \sqrt{\left(\frac{\partial Y}{\partial X_1}\delta X_1\right)^2 + \left(\frac{\partial Y}{\partial X_2}\delta X_2\right)^2 + \cdots + \left(\frac{\partial Y}{\partial X_n}\delta X_n\right)^2} \tag{5.3}$$

δY is the absolute uncertainty of the parameter Y, and δX_1, δX_2,…, δX_n, are the overall uncertainty of parameters X_1, X_2,…, X_n i.e. the directly measurable physical parameters. The maximum relative uncertainty values for the dehumidifier, regenerator and complete system performance studies are presented in their respective experimental results section. Next, Sect. 5.2.3 describes the experimental method.

5.2.3 Experimental Method

The novel IDCS is installed at The University of Nottingham's Marmont Laboratory. This is to facilitate testing under varying environmental and operating conditions in controlled laboratory conditions. There are three main components to the laboratory experimental set-up shown in Fig. 5.7: (1) the novel IDCS, (2) hot water cylinder and (3) environmental chamber. Table 5.2 provides IDCS air flow identification.

The use of the environmental chamber facilitates: (a) a high level of control and provides consistent inlet air conditions to the IDCS throughout all tests, and (b) simulation of different climates other than the UK; specifically those that favour the use of liquid desiccant air conditioning i.e. high humidity. The environmental

Fig. 5.7 IDCS laboratory setup with labelled air flows

Table 5.2 IDCS air flow identification

Air flow ID	Air flow description
A	Regenerator HMX inlet
B	Regenerator HMX outlet
C	Dehumidifier HMX inlet
D	Dehumidifier HMX outlet
E	Evaporative cooler inlet
F	Evaporative cooler outlet

chamber can create air conditions from 0 to 40 °C and 10 to 80 % RH. The dehumidifier (supply) air stream is connected to the environmental chamber by the way of a plenum box. However, the regenerator and evaporative inter-cooler air streams use the air from the laboratory environment. This is because the complete IDCS could not fit in the environmental chamber, and the air flow requirements of the entire IDCS were too high to duct all air flows from the chamber to the IDCS. The desiccant evaporative inter-cooler will perform better with laboratory (room) air as opposed to environment chamber (outside) air, as it is drier, and thus represents a greater evaporative potential. Similarly, the regenerator will perform better with lower humidity laboratory air because it possesses a lower vapour pressure. Therefore using laboratory (room) air for the evaporative and regeneration processes will improve system performance. All outlet air flows are to the laboratory environment.

At the beginning of a test, the temperature and relative humidity of the environmental chamber are set. Depending on the requirements it can take up to 1 h to achieve stable and homogenous air conditions inside. The temperature and relative humidity shown on the chambers display panel is cross checked against the Vaisalia HMP110 humidity and temperature probe at the IDCS dehumidifier inlet and an RS 1365 handheld humidity-temperature meter within the chamber. Once the desired air conditions are achieved, and depending on the test variable under investigation, the IDCS operation is set accordingly and run at that condition.

For desiccant solution regeneration, a vented 120 L hot water cylinder with a 3 kW electrical immersion heater is used as the thermal input. Before the start of a test the hot water tank heater and circulation pump (H/P) are switched on. A by-pass loop is used to circulate the water around the tank until it reaches the desired temperature for the particular test. A control valve (V3) on the flow pipe is used to provide the desired hot water flow to the IDCS. The tank thermostat is set according to the required flow temperature. The flow temperature from the tank is checked at regular intervals, and additional control of the flow temperature is achieved by turning the heater on and off manually.

The desiccant concentration in the desiccant solution tank is recorded at the start, middle and end of each separate test, and the result recorded. The air velocity is measured at each duct outlet and recorded at the beginning of each test, and the result recorded. Multiplication of the average air velocity by the air duct area provides the volumetric air flow through the HMX cores. The desiccant solution and water volumetric flows are measured at the start of a test, and the flow indictors checked periodically throughout a test. Depending on the test variable being investigated, tests last for 30–60 min or until steady-state outlet air conditions are achieved for extended periods (30 min or more). Data is recorded every ten seconds in this period. Only steady-state data is used in the performance evaluation. For each variable investigated there were a minimum of three individual tests conducted. The results presented are the average of each of these tests.

Testing of the dehumidifier and regenerator components is carried out concurrently; this is due to the operational nature of the combined IDCS. The performance evaluation metrics and equations used in the dehumidifier and regenerator tests are the same as those used for the modelling, and can be referred to in Sect. 3.3.3. In the dehumidifier: moisture removal rate, change in absolute humidity of air, dehumidifier cooling output and latent effectiveness are used. In the regenerator: moisture addition rate and regenerator thermal input are used. For the overall system performance, the COP_{th} and COP_{el} are used, and are previously defined in Eqs. 4.30 and 4.31 respectively. The thermophysical properties of the humid air are determined from in-built functions in EES (see Appendix 2 for reference), and the thermophysical properties of the desiccant solution are determined from linear regression curve fits to published data (James 1998; Melinder 2007), presented in Sect. 3.2 and Appendix 1. Uncertainty analysis is used to provide a measure of the experimental error associated with the calculated values, using the propagation of error formula, shown in Eq. 5.3.

Section 5.2.3 has described the experimental method used to test the IDCS. Next, Sect. 5.3 presents the results and analysis from the testing of the IDCS.

5.3 IDCS Results and Analysis

This section presents the results and analysis from the dehumidifier, regenerator and complete IDCS evaluation. Due to the combined nature of the IDCS the desiccant solution flow in the regenerator HMX has to equal that in the dehumidifier HMX. Due to the combined integrated nature of the system, measurement of the desiccant solution properties between the regenerator and dehumidifier is not possible. Unless otherwise varied, Table 5.3 provides the operating values used in the experimental testing of the dehumidifier, regenerator and complete IDCS.

Throughout all tests a desiccant solution volumetric flow of 1.5 L min^{-1} is used. It was found through experimental testing that a volumetric flow above 1.5 L min^{-1} resulted in desiccant solution entrainment in the supply airstream, and below 1.5 L min^{-1} leads to insufficient wetting of the membrane surface.

Next, Sect. 5.3.1 presents the results and analysis from the IDCS dehumidifier component evaluation.

Table 5.3 Operating values in the evaluation of the IDCS

Variable	Dehumidifier	Inter-cooler	Regenerator
Desiccant/water flow (L min^{-1})	1.5	1.5	1.5
Desiccant temperature (°C)	23–26	–	–
Water temperature (°C)	–	22–25	–
Solution mass concentration (–)	0.65–0.7	–	0.65–0.7
Volumetric air flow (m^3 h^{-1})	243	269	243
Inlet air temperature (°C)	30	22–26	22–26
Inlet air relative humidity (%)	60	38–66	38–66

5.3.1 IDCS Dehumidifier Component Analysis

The role of the dehumidifier is to cool a supply air stream through the lowering of its enthalpy. Enthalpy reduction is achieved primarily through the removal of moisture from the air stream to a liquid desiccant solution. Depending on the desiccant solution temperature, a reduction in the supply air temperate may also occur.

This section presents the results and analysis from the IDCS dehumidifier component evaluation. The IDCS dehumidifier component evaluation has assessed the impact; inlet air temperature, inlet air relative humidity and volumetric air flow has on dehumidifier performance. The key dehumidifier performance metrics considered in this section are: moisture removal rate, change in absolute humidity of the air across the dehumidifier, latent (dehumidifier) effectiveness and dehumidifier cooling output. Table 5.3 lists the operating values used in the dehumidifier testing.

5.3.1.1 IDCS Dehumidifier Inlet Air Condition Effect

The IDCS dehumidifier performance has been evaluated over a 50–70 % relative humidity range at a 30 and 35 °C inlet air temperature. The data presented in Fig. 5.8 shows that dehumidifier performance improves with increasing inlet air temperature and relative humidity. Figure 5.8a shows that over the investigated relative humidity range the moisture removal rate from the supply airstream increases from 0.1541 to 0.4395 g s^{-1} for the 30 °C inlet air condition and 0.2354–0.4682 g s^{-1} for the 35 °C inlet air condition. As the relative humidity and temperature of the inlet air increases, its vapour pressure increases, and thus a greater vapour pressure difference between the humid air and desiccant solution exists, driving greater mass transfer. Figure 5.8b shows that over the investigated relative humidity range the absolute humidity difference of the supply air stream increases, i.e. more dehumidification occurring, from 0.001988 to 0.005728 kg_{vapour}/kg_{dryair} for the 30 °C inlet air condition, and from 0.003073 to 0.0062 kg_{vapour}/kg_{dryair} for the 35 °C inlet air condition. Figure 5.8c shows that over the investigated relative humidity range the latent (dehumidifier) effectiveness increases from 29.91 to 38.39 % for the 30 °C inlet air condition, and from 32.32 to 46.78 % for the 35 °C inlet air condition. Figure 5.8d shows that over the investigated relative humidity range the cooling output from the dehumidifier increases as the inlet air relative humidity and temperature increases. The dehumidifier cooling ranges from 570 to 1084 W at an inlet temperature of 30 °C, and from 1059 to 1362 W at an inlet temperature of 35 °C. The increase in cooling output with air relative humidity and temperature is due to greater moisture removal rate and thus greater latent cooling achieved plus a greater temperature difference between the air and desiccant solution, leading to increased sensible cooling.

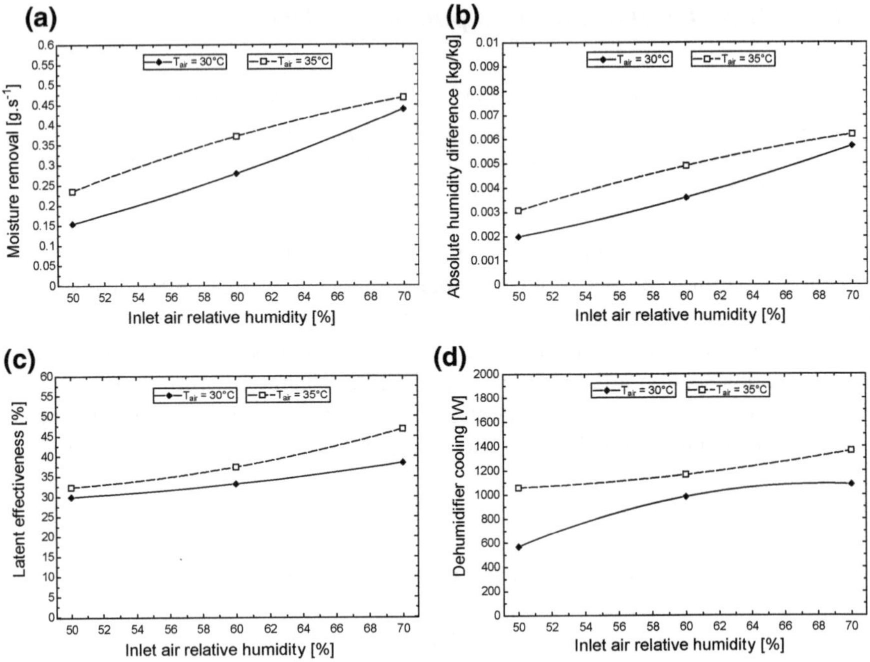

Fig. 5.8 IDCS dehumidifier performance with inlet air conditions

At the 30 and 35 °C inlet air condition, the average supply air temperatures across all relative humidity tests is 28.81 and 31.97 °C respectively. From Fig. 5.8 it is evident that the IDCS dehumidifier performance improves with an increase in inlet air temperature and relative humidity. The system is therefore well suited to hotter, more humid climate such as Southeast Asia. However increased performance will result in a greater dilution of the desiccant solution. For tri-generation system integration, consideration needs to be given to whether the regenerator moisture addition rate achievable with the SOFC thermal output can match the mass removal rate in the dehumidifier.

The IDCS evaporative inter-cooler is included to enhance performance by providing sensible cooling to the dehumidification process. The evaporative inter-cooler is operated on laboratory air. Figure 5.9 shows the relationship between the inlet air absolute humidity to the evaporative inter-cooler and the cooling effect it provides. The cooling output is determined based on the enthalpy difference of the inter-cooler's inlet and outlet air. At an inlet air condition of 0.007 kg_{vapour}/kg_{dryair} around 800 W of cooling is achieved, this reduces to around 400 W at a 0.011 kg_{vapour}/kg_{dryair} inlet air condition. At lower inlet air absolute humidity values, the evaporative cooler produces a greater cooling output due to the inlet air having a lower wet-bulb temperature and thus greater evaporative potential. As a result, in a building application, it would be recommended to operate the evaporative inter-cooler on drier return room air, as opposed to fresh outside (humid) air, to increase cooling output.

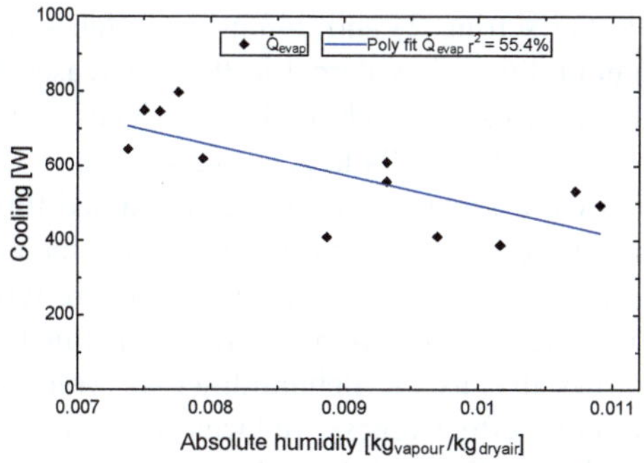

Fig. 5.9 IDCS evaporative-inter cooler output with inlet air absolute humidity

5.3.1.2 IDCS Dehumidifier Volumetric Air Flow Effect

Figure 5.10 shows the impact inlet air volumetric flow has on dehumidifier performance at a set inlet condition of 30 °C and 60 % relative humidity. Figure 5.10a shows the moisture removal rate increases with volumetric air flow, from 0.2058 g s^{-1} at 124 m^3 h^{-1} to a maximum of 0.2978 g s^{-1} at 243 m^3 h^{-1}. There is little difference (<0.0116 g s^{-1}) between the moisture removal rate achieved

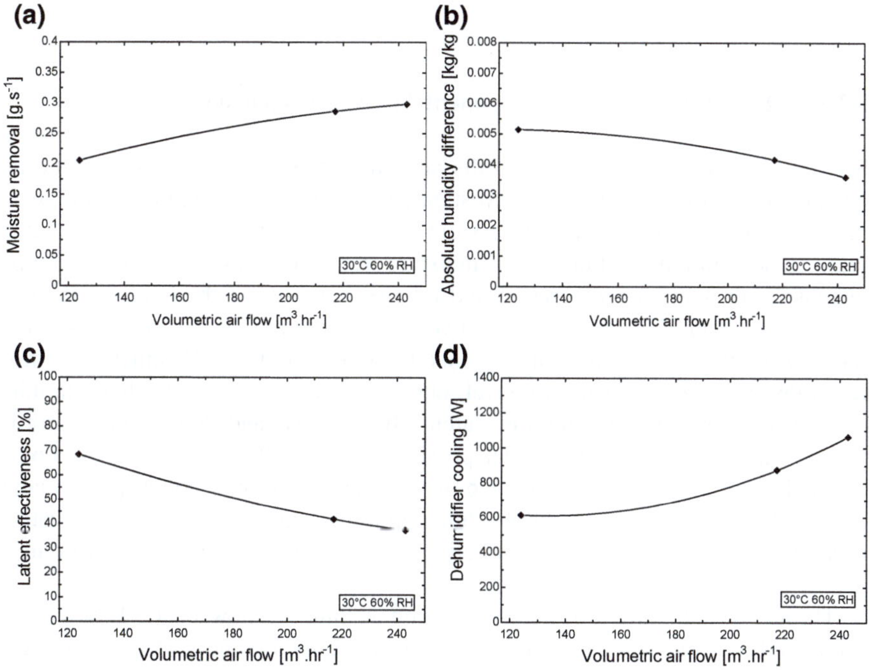

Fig. 5.10 IDCS dehumidifier performance with inlet air volumetric flow

between 217 $m^3 h^{-1}$ (fan setting 2) and 243 $m^3 h^{-1}$ (fan setting 3). Figure 5.10b shows that as the dehumidifier air volumetric flow increases, the change in absolute humidity of the air across the dehumidifier reduces from 0.005146 kg_{vapour}/kg_{dryair} at 124 $m^3 h^{-1}$ to 0.003594 kg_{vapour}/kg_{dryair} at 243 $m^3 h^{-1}$. As volumetric air flow increases, a greater mass of air is passed through the dehumidifier, and thus the capacity of the dehumidifier to reduce the air absolute humidity reduces. This relationship is in conflict with the moisture removal rate shown in Fig. 5.10a; this is because moisture removal rate is a function of air mass flow rate. Figure 5.10c shows that as the dehumidifier air volumetric flow increases, the latent (dehumidifier) effectiveness reduces from 68.52 % at 124 $m^3 h^{-1}$ to 37.35 % at 243 $m^3 h^{-1}$. Figure 5.10d shows the dehumidifier cooling output increases as the volumetric air flow increases, from a minimum of 613 W at 124 $m^3 h^{-1}$ to 1065 W at 243 $m^3 h^{-1}$. This is primarily due to a larger volume of air being conditioned.

The selection of an appropriate volumetric air flow in the dehumidifier is dependent upon the application and the desired supply air condition. However, in a tri-generation system it will be advantageous to operate the dehumidifier at the highest volumetric air flow in order to maximise the cooling output, and thus tri-generation efficiency. Across all dehumidifier tests, the maximum calculated relative uncertainties in the dehumidifier MRR, $\Delta\omega$, η_L and $\dot{Q}_{cooling}$ are ±15.98, ±15.1, ±12.47, and ±15.04 % respectively.

Next, Sect. 5.3.1.3 provides the dehumidifier component evaluation conclusions.

5.3.1.3 IDCS Dehumidifier Component Analysis Conclusions

Section 5.3.1 has presented the results and analysis from the IDCS dehumidifier component evaluation. Over the investigated environmental conditions, the dehumidifier performs well with a potassium formate solution at a 0.65–0.7 solution mass concentration. Dehumidifier moisture removal rates and cooling output increase with inlet air temperature and relative humidity in the range of 0.1541–0.4682 $g s^{-1}$ and 570–1362 W respectively. The dehumidifier effectiveness values range from 30 to 47 %, typical of a membrane based HMX. Volumetric air flow has little impact on moisture removal, but a marked impact on absolute humidity difference across the dehumidifier, latent effectiveness and dehumidifier cooling output. The evaporative inter-cooler provides between 400 and 800 W of cooling to the dehumidifier. The performance of the evaporative-inter cooler performance is strongly linked to the inlet air absolute humidity. Thus, in a building application, it is beneficial to operate the evaporative inter-cooler on drier return room air to maximise performance.

Throughout all tests a desiccant solution volumetric flow of 1.5 $L min^{-1}$ has been used as it provides sufficient wetting of the membrane surface area whilst (initially) minimising solution leakage into the supply air stream. However, as shown in Fig. 5.11 desiccant solution leakage has occurred in the IDCS design.

Fig. 5.11 Dehumidifier HMX leakage in the IDCS

Resolving the leakage problem is essential for effective IDCS tri-generation system building integration and to avoid the issues of: poor indoor air quality, desiccant losses and duct corrosion.

Next, Sect. 5.3.2 presents the results and analysis from the IDCS regenerator component evaluation.

5.3.2 IDCS Regenerator Component Analysis

The aim of the regeneration process is to remove the water vapour gained by the desiccant solution during the dehumidification process. The moisture removal from the dehumidifier air stream to the desiccant solution should equal the moisture addition from the desiccant solution to the regeneration air stream and thus the complete system can run continuously. Table 5.3 lists the operating values used in the regenerator testing. In the experimental work presented, ambient laboratory conditions have been used as the regenerator inlet air stream.

This section presents the results and analysis from the IDCS regenerator component evaluation. The IDCS regenerator evaluation has assessed the impact inlet air absolute humidity, volumetric air flow and volumetric water flow in the heating circuit have on regenerator performance. The key regenerator performance metrics considered in this section are: moisture addition rate and regenerator thermal input. The regenerator thermal input, $\dot{Q}_{reg}$ in Watts is determined using Eq. 5.4.

$$\dot{Q}_{reg} = \dot{m}_w c_{p_w} \left(T_{w,\text{flow}} - T_{w,\text{return}} \right) \tag{5.4}$$

$\dot{m}_w$ and $c_{p,w}$ are the mass flow rate (kg s^{-1}) and specific heat capacity (J kg^{-1} K) of the water in the heating circuit respectively. $T_{\text{w,flow}}$ and $T_{\text{w,return}}$ are the respective flow and return temperatures (°C) of the water in the heating circuit. During

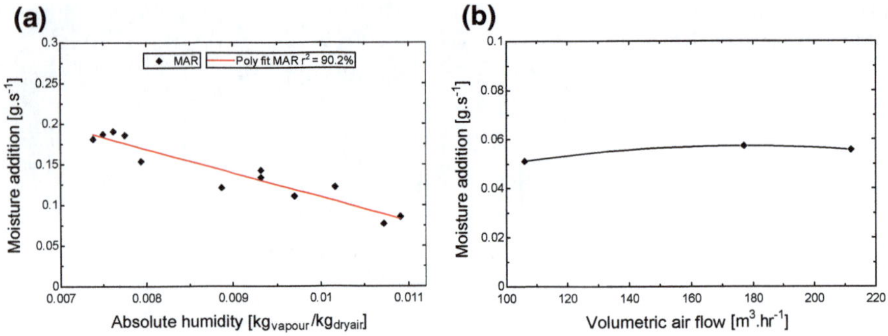

Fig. 5.12 IDCS regenerator performance with **a** inlet air absolute humidity, and **b** volumetric air flow

regenerator testing a water flow temperature of 60 °C and water volumetric flow in the heating circuit of 2 L min^{-1} is used. The thermal input values used in the regenerator evaluation are based upon those typical of a SOFC CHP system found in the literature, such as the BlueGEN unit presented in Chap. 7. This assumption aims to simulate the IDCS operation in a SOFC tri-generation system application. Figure 5.12a shows the impact inlet air absolute humidity to the regenerator has on moisture addition rate. The inlet air temperature to the regenerator ranges from 22 to 26 °C, and the absolute humidity ranges from 0.00708 to 0.01197 kg_{vapour}/kg_{dryair}. The moisture addition rate ranges from a minimum of 0.07715 g s^{-1} to a maximum of 0.2229 g s^{-1}. Mass transfer is driven by a vapour pressure difference between the desiccant solution and the regenerator airstream. As the absolute humidity of the regenerator inlet airstream increases so does its vapour pressure, resulting in a smaller moisture addition rate.

Figure 5.12b shows the variation of moisture addition rate from the desiccant solution to the regenerator airstream as a function of regenerator volumetric air flow. The novel IDCS integrates three components; dehumidifier, regenerator and evaporative cooler into one HMX core. As a result, the operation of each component has an impact on the others. During volumetric air flow testing, the regenerator was operated independently i.e. no dehumidifier or evaporative cooler, and as a result the regenerator volumetric air flow shown in Fig. 5.12b is lower than that observed during simultaneous dehumidifier and regenerator operation. The regenerator volumetric air flow rate is increased from 106 to 212 m^3 h^{-1}. It is evident that the volumetric air flow has little impact on the moisture addition rate, with values ranging between 0.05118 and 0.05727 g s^{-1}, an increase of 0.00609 g s^{-1}.

The IDCS regenerator performance has been evaluated with respect to the volumetric water flow in the regenerator hot water heating circuit, over a 1.5–6.5 L min^{-1} range. Figure 5.13a shows the variation of the moisture addition rate in the regenerator with respect to the volumetric water flow. It is evident that the volumetric flow of the water has a marginal impact of the regenerator capacity, with the moisture addition rate ranging from 0.2363 to 0.2619 g s^{-1}, a difference of 0.0256 g s^{-1}.

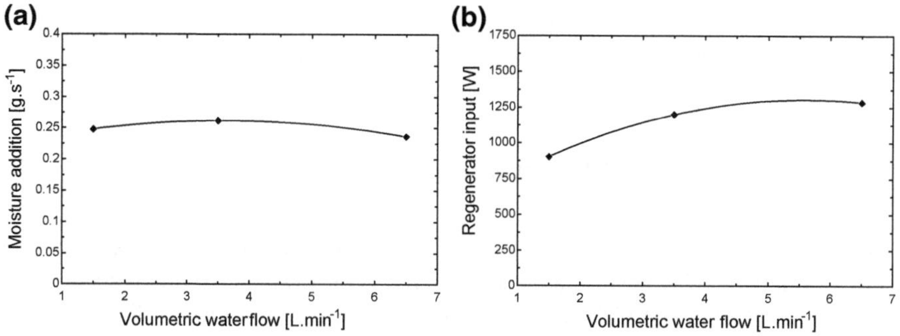

Fig. 5.13 IDCS regenerator performance with heat exchanger volumetric water flow

Figure 5.13b shows the regenerator thermal input as a function of volumetric water flow in the regenerator liquid to air heat exchanger. The volumetric water flow has a large impact on the thermal input to the system. At 1.5 L min^{-1} the thermal input is 903 W at 6.5 L min^{-1} the thermal input is 1285 W. As highlighted in Fig. 5.13a the volumetric water flow has little or no impact on the moisture addition rate in the regenerator, it is therefore optimal to operate the IDCS at a 1.5 L min^{-1} volumetric water flow in the regenerator hot water circuit. As discussed in Sect. 5.3.3, having a lower regenerator thermal input but a similar moisture addition rate will assist in improving the COP_{th} of the total system. Across all regenerator tests, the maximum calculated relative uncertainty in the regenerator MAR is ±25.6 %.

Next, Sect. 5.3.2.1 provides the regenerator component evaluation conclusions.

5.3.2.1 IDCS Regenerator Component Analysis Conclusions

Section 5.3.2 has presented the results and analysis from the IDCS regenerator component evaluation. Regeneration capacity increases with a lower inlet air absolute humidity. As a result it is recommended to operate the regenerator on drier return room air in a building application. Volumetric air flow and volumetric water flow in the heating circuit has marginal impact on regenerator capacity in the IDCS design. However, the volumetric water flow does influence the regenerator thermal input and should be minimised. It is evident, across the conditions investigated, that there is an issue of instantaneous mass imbalance between the dehumidifier and regenerator i.e. the mass of water vapour removed from the air in the dehumidifier does not equal the mass removed from the desiccant solution in the regenerator. As a result, the complete IDCS cannot run continuously because the solution will become weak over time. The mass imbalance issue is discussed in more detail in the complete system analysis and chapter conclusions in Sects. 5.3.3 and 5.4 respectively.

Next, Sect. 5.3.3 presents the results and analysis from the complete IDCS component evaluation.

5.3.3 Complete IDCS Performance Analysis

This section presents the results and analysis from the complete IDCS component evaluation. The performance of the IDCS is evaluated with respect to its COP_{th} and COP_{el}. The COP equations are previously presented in Eqs. 4.30 and 4.31 respectively. An issue encountered with the IDCS is that an instantaneous mass balance between the dehumidifier and regenerator is not easily achievable. Mass imbalance is primarily due to the available surface area for heat and mass exchange in the regenerator being too small and an insufficient vapour pressure differential between the air and desiccant solution.

In order to regenerate the desiccant solution back to its original condition following the dehumidification process, the regenerator needs to operate for extended periods. As a result of this imbalance, a theoretical adjusted thermal COP, $COP_{th,adj}$, has been proposed in Eq. 5.5. The $COP_{th,adj}$ is a steady state value that takes into account the requirement of extended regenerator operation in order to achieve a mass balance.

$$COP_{th,adj} = \frac{\dot{Q}_{cooling}}{\left(\frac{MRR}{MAR}\right)\dot{Q}_{reg}} \tag{5.5}$$

Figure 5.14a shows the average $COP_{th,adj}$ and COP_{el} for 21 IDCS tests. The inlet air conditions for each of these tests can be referred to in Appendix C. The black horizontal lines at $y = 1$ and $y = 2$ mark the benchmark values for COP_{th} and COP_{el} respectively. The $COP_{th,adj}$ values range from a minimum of 0.34 to a maximum of 1.26, with an average of 0.72. A COP_{th} above 1.0 is comparable with 5.0 for a VCS driven by grid electricity, demonstrating the potential for highly

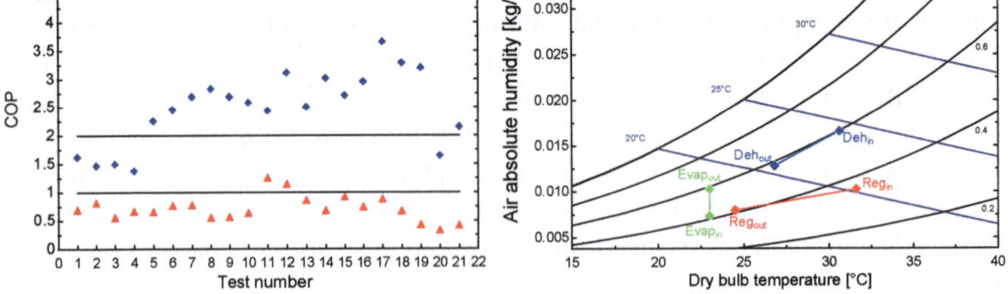

Fig. 5.14 **a** Complete IDCS performance, and **b** IDCS psychrometric process

efficient air conditioning with the IDCS design. Furthermore, the $COP_{th,adj}$ values are competitive with current VAS, but at a smaller cooling capacity, highlighting the potential for liquid desiccant air conditioning technology in the development of an efficient and effective tri-generation system for (domestic) building applications. The COP_{el} values range from a minimum of 1.38 to a maximum of 3.67, with an average of 2.49. Figure 5.14b demonstrates the psychrometric process of the complete IDCS, indicating the air state points in the dehumidifier, regenerator and evaporative inter-cooler. The data points in Fig. 5.14b are taken from test seven in Fig. 5.14a.

As the moisture addition rate in the regenerator increases, the $COP_{th,adj}$ increases. This is due to an improved mass balance between the dehumidifier and regenerator, and as a result a lower adjusted regenerator thermal input. The two $COP_{th,adj}$ values greater than 1.0 are attained when the moisture addition rate in the regenerator is greater than 0.17 g s^{-1}, which is achieved when the absolute humidity of the inlet air to the regenerator is less than $0.008 kg_{vapour}/kg_{dryair}$. Thus, it can be concluded that the IDCS performs best when moisture addition in the regenerator is maximised, which occurs at a lower regenerator inlet air absolute humidity value. As a result, when operating the regenerator on fresh outside air, liquid desiccant system performance will be poorer in hot and high humid climates. The reasonable $COP_{th,adj}$ values demonstrate the potential of liquid desiccant air conditioning technology in the development of an efficient and effective tri-generation system for (domestic) building applications. The maximum calculated relative uncertainties in the IDCS $COP_{th,adj}$ and COP_{el} are ± 27.73 and ± 15.93 % respectively.

Next, Sect. 5.4 presents the chapter conclusions and the potential of the IDCS for successful tri-generation system integration.

5.4 Chapter Conclusions and IDCS Potential for Tri-generation System Integration

To date, the application of liquid desiccant air conditioning in tri-generation (domestic) built environment applications has been limited. This is primarily due to three main issues: large system size, system complexity and issues of desiccant solution leakage/carry-over. As a result, a novel IDCS has been developed with the aim of overcoming these barriers in order to facilitate effective tri-generation system integration. The IDCS combines the regenerator, evaporative inter-cooler and dehumidifier into a single HMX core to reduce size and limit the amount of piping, heat exchangers and pumps. A semi permeable micro-porous membrane is used to prevent solution entrainment in the supply air stream. No previous work has been found in the literature regarding such an integrated design, and thus the work presented is a clear contribution to knowledge, providing progress to the fields of liquid desiccant air conditioning technology for building and tri-generation system applications.

This chapter has presented a component evaluation, based on experimental data, of the novel IDCS design. An environmentally friendly potassium formate working fluid has been employed at a solution mass concentration of 0.65–0.7. The IDCS experimental set-up, instrumentation and experimental method have all been provided. The results and analysis have been presented in three sections: dehumidifier, regenerator and complete IDCS. The aim of the experimental IDCS component testing is to: (1) highlight the impact changes in environmental and operating conditions have on performance, (2) record the operating conditions in which the dehumidifier and regenerator are balanced, and (3) discuss the potential for tri-generation system integration, and the operating conditions to facilitate effective coupling.

Over the investigated environmental and operating conditions, the dehumidifier performs well with a potassium formate solution. Dehumidification capacity increases with inlet air temperature, relative humidity and air volumetric flow. Issues of desiccant solution leakage and entrainment in the supply air stream have been encountered; however improved manufacturing and sealing will address this problem. Using thermal input values based upon those found in a typical SOFC CHP WHR circuit regeneration of the potassium formate solution is possible, demonstrating the potential for the development of a tri-generation system based on SOFC and liquid desiccant air conditioning technology. However, a significant conclusion from the work presented is that an instantaneous mass balance between the dehumidifier and regenerator is challenging under most conditions. Across the variables investigated there is a greater instantaneous moisture removal rate in the dehumidifier than moisture addition rate in the regenerator.

As a result, a theoretical adjusted COP_{th} ($COP_{th,adj}$) has been presented which takes into account the requirement of extended regenerator operation in order to achieve a mass balance. The IDCS performs best when moisture addition in the regenerator is maximised, which occurs at a lower regenerator inlet air absolute humidity value. Across all tests performed, an average $COP_{th,adj}$ of 0.72 has been achieved.

The component testing presented in this chapter has demonstrated that the novel IDCS concept is viable and has provided progress to the field of liquid desiccant air conditioning technology. The dehumidifier performs well with a potassium formate solution across the parameters investigated. However, three key operational issues mean the IDCS is not considered suitable for tri-generation system integration, these are as follows:

1. Mass imbalance between dehumidifier and regenerator, and thus extended regenerator operation required.
2. Desiccant solution leakage/carry-over on the dehumidifier side of the HMX.
3. As highlighted in Chap. 4, it is primarily the optimisation of the liquid desiccant component that facilitates effective tri-generation system integration and operation. With limited control in the IDCS operating variables, tri-generation system integration will be challenging.

Future IDCS work should focus on increasing the regenerator to dehumidifier HMX surface area ratio and improving heat transfer rates to the regenerator air stream to improve mass balance. Better sealing of the membrane HMX will prevent (limit) solution leakage/carry-over. This would also mean a higher solution volumetric flow may be permissible, thus increasing the IDCS dehumidification capacity.

In response to the highlighted shortcomings of the novel IDCS a separate liquid desiccant air conditioning system (SDCS) has been developed for tri-generation system integration. Chapter 6 presents the evaluation, based on experimental data, of the SDCS.

References

Beccali, M., P. Finocchiaro, and B. Nocke. 2012. Energy performance evaluation of a demo solar desiccant cooling system with heat recovery for the regeneration of the adsorption material. *Renewable Energy* 44: 40–52.

Elmer, T., M. Worall, S. Wu, and S. Riffat. 2016c. An experimental study of a novel integrated desiccant air conditioning system for building applications. *Energy and Buildings* 111: 434–445.

Gommed, K., and G. Grossman. 2007. Experimental investigation of a liquid desiccant system for solar cooling and dehumidification. *Solar Energy* 81(1): 131–138.

Halliday, S.P., C.B. Beggs, and P.A. Sleigh. 2002. The use of solar desiccant cooling in the UK: A feasibility study. *Applied Thermal Engineering* 22(12): 1327–1338.

Hassan, H.Z., and A.A. Mohamad. 2012. A review on solar cold production through absorption technology. *Renewable and Sustainable Energy Reviews* 16(7): 5331–5348.

Jain, S., and P.K. Bansal. 2007. Performance analysis of liquid desiccant dehumidification systems. *International Journal of Refrigeration* 30(5): 861–872.

James, S. 1998. A new working fluid 'potassium formate' for use in absorption heat pumps. Masters of Philosophy, The University of Nottingham.

Liu, S. 2008. A novel heat recovery/desiccant cooling system. PhD, The University of Nottingham.

Melinder, A. 2007. Thermophysical properties of aqueous solutions used as secondary working fluids. PhD Thesis, KTH Energy and Environmental Technology.

Taylor, J.R. 1997. *An introduction to error analysis: The study of uncertainties in physical measurements*. Sausalito, CA: University Science Books.

Chapter 6
Component Evaluation of a Separate Liquid Desiccant Air Conditioning System (SDCS)

6.1 Introduction

This chapter aims to evaluate the performance of a separate liquid desiccant air conditioning system (SDCS) under a variety of environmental and operating conditions. SDCS component evaluation facilitates effective tri-generation system integration presented in Chap. 7.

Conclusions drawn from Chap. 5 demonstrate that the developed novel IDCS would not be suitable in a SOFC tri-generation system, particularly for building applications. This is due to a mass imbalance between the dehumidifier and regenerator, desiccant solution leakage and poor controllability of operating variables. Unlike the IDCS, the SDCS consists of three separate cores: a dehumidifier, regenerator and evaporative cooler. The SDCS has been developed in response to the IDCS shortcomings and with the intention that it will demonstrate greater potential for tri-generation system integration. The SDCS has been acquired from the company Intelligent Integration of Solar Air and Water (IISAW) in China, but has been tested at The University of Nottingham. The experimental work involves the testing of the heat powered SDCS under varying environmental and operating conditions in controlled laboratory conditions, using a potassium formate ($CHKO_2$) liquid desiccant solution. As will be demonstrated in Chap. 7, there is limited control of the outputs for the SOFC CHP unit. As a result it is the operation of the SDCS that will need to be optimised to facilitate successful pairing of the two systems into an efficient and effective tri-generation system.

Detailed component testing of the SDCS under controlled laboratory conditions will facilitate the following:

(1) Highlight the impact changes in environmental and operational conditions have on the performance of the dehumidifier, regenerator and complete SDCS.
(2) Recording of the operational conditions in which the dehumidifier and regenerator are balanced i.e. the mass of water absorbed in the dehumidifier

T. Elmer, *A Novel SOFC Tri-generation System for Building Applications*,
Springer Theses, DOI 10.1007/978-3-319-46966-9_6

is equal to the mass removed in the regenerator. This will allow continuous operation of the SDCS across a range of working requirements.

(3) Selection of optimised operational conditions for the SDCS, with particular consideration to tri-generation system integration i.e. desiccant solution volumetric flow and hot water volumetric flow in the heating circuit (SOFC).

The chapter is split into three sections. Section 6.2 provides details of the SDCS experimental set-up, instrumentation and experimental method. Section 6.3 presents the SDCS experimental results and analysis. Section 6.4 concludes the chapter and discusses the implications for tri-generation system integration in an experimental setting. A published journal article, by the author, on the component evaluation of the separate liquid desiccant air conditioning system can be referred to here; Elmer et al. (2016b).

6.2 SDCS Experimental Set-up

This section provides a detailed description of the experimental set-up, instrumentation and experimental method used in the testing of the SDCS.

As previously discussed in Sect. 5.4, operational issues with the IDCS means it is not suitable for tri-generation system integration. Therefore a SDCS has been developed. As with the IDCS, the SDCS uses a semi-permeable micro porous membrane based cross flow contactor, operating with a low cost, environmentally friendly, non-corrosive potassium formate desiccant solution. Figure 6.1 shows a schematic diagram of the complete SDCS with labelled components.

The SDCS consists of three separate cores; dehumidifier (D/C), regenerator (R/C) and evaporative cooler (E/C). In the dehumidifier and regenerator, each HMX is 410 mm in length, 210 mm in height and 230 mm in width. The cores consist of 21 channels that allow air and desiccant solution to flow in a cross flow manner (air through the core, desiccant downwards through the core), separated by a fibre membrane. The solution channels consist of a polyethylene sheet, with fibre membranes attached on either side. The gap between the two solution channels provide the space for the air to flow. The semi permeable micro-porous fibre membrane allows the diffusion of water vapour, but prevents liquid desiccant solution migrating across it, thus overcoming the issue of liquid desiccant entrainment in the air stream. The evaporative cooler core is also 410 mm in length, 210 mm in height and 230 mm in width, with 21 air channels. There is no inclusion of a membrane, only a fibre material to create the air channels and provide a wetted surface for the water to flow down. Air and water come into contact in a cross flow manner. All three cores are housed in a polyethylene box as shown in Fig. 6.2.

The dehumidification system consists of a 15 L polyethylene tank and a 15 W single phase centrifugal magnetically driven pump (0–10 L min^{-1}), which delivers strong desiccant solution to the top of the dehumidifier membrane HMX through a spray nozzle. Prior to the desiccant solution entering the dehumidifier membrane

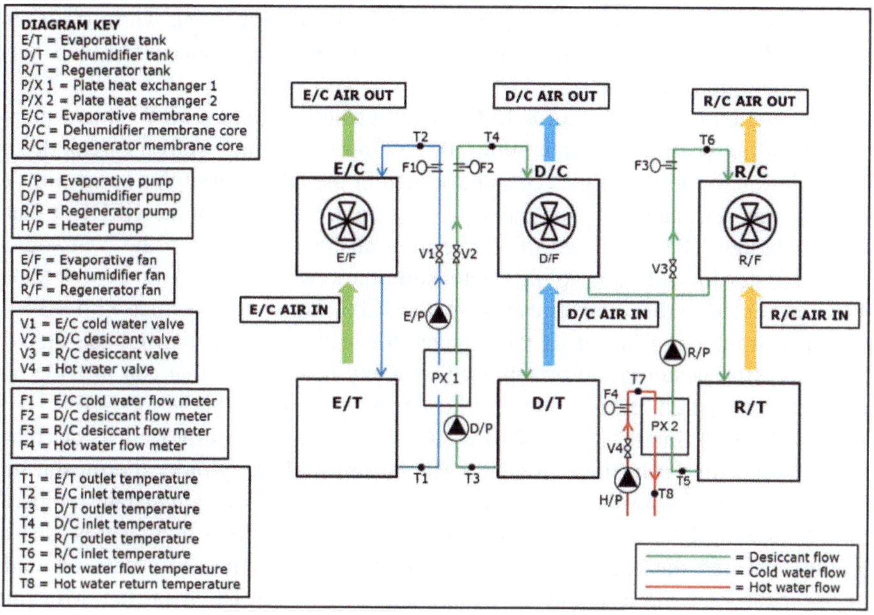

Fig. 6.1 SDCS schematic with labelled components

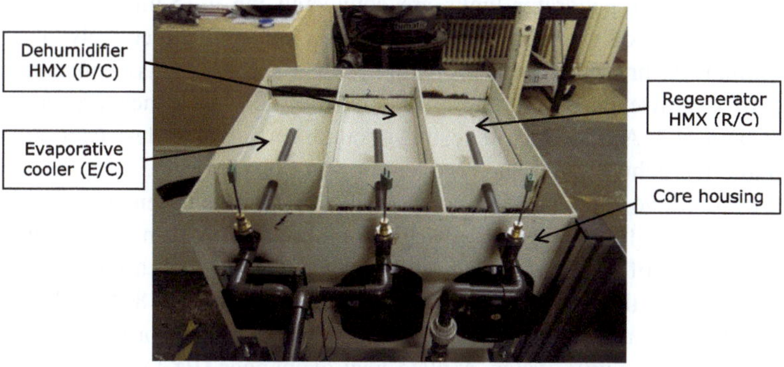

Fig. 6.2 SDCS HMX cores

HMX, a plate heat exchanger (PX1) is used to pre-cool the desiccant solution using cooling water from the evaporative cooler. This serves to increase the dehumidification potential of the solution and provide sensible cooling to the supply air. The strong cool desiccant solution flows downwards, due to gravity, through the dehumidifier membrane HMX. Humid air is passed across one side of the membrane via a 400 $m^3\ h^{-1}$ (nominal) 23 W 24 V DC axial fan. The supply air is dehumidified by absorption of moisture into the desiccant solution, and depending on the solution temperature the air is sensibly cooled. The weak warm desiccant solution flows out of the bottom of the dehumidifier membrane HMX and back into the dehumidifier tank.

The regenerator consists of a 15 L polyethylene tank and a 15 W single phase centrifugal magnetically driven pump (0–10 L min^{-1}), which delivers weak desiccant solution to the top of the regenerator membrane HMX through a spray nozzle. Prior to the desiccant solution entering the regenerator membrane HMX, a plate heat exchanger (PX2) is used to enable the transfer of heat from the heat source to the desiccant solution to facilitate solution regeneration. The experimental work presented in this chapter uses a vented 120 L hot water cylinder with a 3 kW electrical immersion heater as the regenerator heat source. However, the electrical immersion heater could be replaced with any heat source that can provide hot water at the desired temperature and flow rate. In Chap. 7 this is the SOFC CHP system. A Wilo-Smart A-rated 230 V AC pump has been employed to circulate the hot water in the heating circuit. A Honeywell L641A cylinder thermostat has been used to maintain the flow temperature from the tank at a constant temperature. Following PX2 the heated weak desiccant solution flows downwards, due to gravity, through the regenerator membrane HMX. Air is passed across one side of the membrane via a 400 m^3 h^{-1} (nominal) 23 W 24 V DC axial fan. Moisture is vaporised from the desiccant solution and is absorbed by the air, resulting in re-concentration of the desiccant solution. The strong hot desiccant solution flows out of the bottom of the regenerator membrane HMX and back into the tank. The dehumidifier and regenerator tanks are connected via a 15 mm balance pipe to allow the transfer of weak and strong desiccant solution respectively. The potassium formate is provided from the supplier at a solution mass concentration of around 0.74. However, during operation, the solution mass concentration is measured at 0.65–0.7. At the start of testing around ten litres of desiccant solution is loaded into the dehumidifier and regenerator tanks respectively.

The evaporative cooler system consists of a 15 L polyethylene tank and a 15 W single phase centrifugal magnetically driven pump (0–10 L min^{-1}), which delivers water to the top of the evaporative cooler exchanger through a spray nozzle. Air is passed through the exchanger via a 320 m^3 h^{-1} (nominal) 18 W 24 V DC axial fan. Direct evaporative cooling takes place as moisture is absorbed by the air and directly cools the supply water. A plate heat exchanger (PX1) is used to transfer this coolth to the desiccant solution. Figure 6.3 provides a labelled photograph of the complete SDCS.

Next, Sect. 6.2.1 describes the instrumentation used in the experimental testing of the SDCS.

6.2.1 Instrumentation

This section will describe the instrumentation used on the SDCS. Reference to the SDCS schematic in Fig. 6.1 is made throughout. All fans on the SDCS are

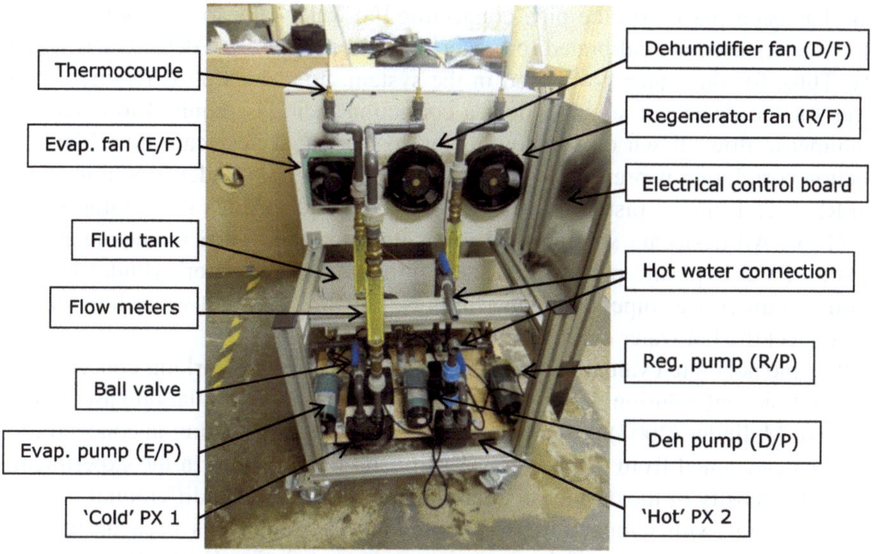

Fig. 6.3 SDCS photograph with labelled components

equipped with variable fan speed controllers to enable control of the volumetric air flow through the cores. The air outlets of the evaporative, dehumidifier and regenerator cores are fitted with 125 mm galvanised steel spiral tube ducting, 500 mm in length. The inlet and outlet air flows have been instrumented with Vaisalia HMP110 humidity and temperature probes. The probes are mounted within the spiral tube ducting using special flanges. The humidity and temperature probes are factory calibrated. Air velocity through the dehumidifier, regenerator and evaporative cooler cores are measured using an RS AM4204 hot wire anemometer at the air ducting outlets. The hot wire anemometer is factory calibrated. Air velocity measurements are recorded at five points across the air duct, and the average taken. The air velocity measurements are validated against a TSI LCA501 rotating vane anemometer.

All liquid flows between the tank and core inlets have been equipped with ball valves (V1–V3) so that the desiccant or water volumetric flow may be set to a desired value. A valve has also been placed on the hot water circuit (V4). All water and desiccant solution flows have been instrumented with sheathed K-Type thermocouples (Nickel Chromium/Nickel Aluminium). Thermocouples have been placed at the inlet to the evaporative, dehumidifier and regenerator cores (T2, T4 and T6) and at the outlets of the evaporative, dehumidifier and regenerator tanks (T1, T3 and T5). Thermocouples have also been placed at the hot water inlet (T7) and outlet (T8) to PX2. The desiccant solution volumetric flow is measured using a 1.5 to 10L.min^{-1} Parker Liquid Flow Indicator; these are placed on the pipe connecting the tank to the core (dehumidifier—F2 and regenerator—F3). The water volumetric flow is measured using a 0.2–2 L min^{-1} Parker Liquid Flow Indicator, and has been placed on the pipe connecting the water tank to the core (F1). The flow meters used are calibrated for water at 20 °C according to density and viscosity. Thus, for the water flows used in the system, no correction is required, however for the desiccant solution flow a correction factor is required to equate the volumetric flow shown on the flow meter, to the actual desiccant flow. This correction correlation is presented in Eq. 5.1. The hot water cylinder is equipped with an RS 1–15 L min^{-1} piston flow meter (F4), designed for flow temperatures of up to 60 °C. All desiccant solution and water flows on the SDCS are equipped with 20 mm PVC-U plastic pipe, with plastic fittings. The hot water cylinder is piped with 22 mm copper pipe and copper fittings. Flexible PVC hot water hose is used to connect the hot water cylinder to PX2 on the SDCS.

For the accurate evaluation of the desiccant system, the working concentration of the desiccant solution needs to be determined. Using a correlation based on the work of Melinder (2007) presented in Eq. 5.2, the desiccant solution concentration can be determined from the solution density and temperature. In the experimental work the density of the desiccant solution is measured using a differential pressure density meter with temperature compensation. The meter has been designed to work in the density range of the potassium formate solution (1400–1550 kg m^{-3}), and has been calibrated with water. The measurement prongs of the differential pressure density meter are placed in the desiccant solution tank and held until a steady state reading is achieved. The temperature of the solution is measured using the K-Type thermocouple at the tank outlet. The concentration is then calculated.

The electrical consumption of fans and pumps is measured using a Brennenstuhl PM230 electricity monitor. This is essential for the COP_{el} calculations. At full load the desiccant system parasitic electrical load is measured at 110 W. The electrical power requirement is measured and recorded for each test. A DataTaker DT500 datalogger is used to record data from the humidity and temperature probes and thermocouples every ten seconds. Further details of the instrumentation and their associated accuracy can be referred to in Table 5.1.

Section 6.2.1 has described the SDCS instrumentation. Next, Sect. 6.2.2 describes the experimental method.

6.2.2 Experimental Method

This section will present the experimental method used in the testing of the dehumidifier, regenerator and complete SDCS. In the laboratory testing of the SDCS, an environmental chamber has been employed to (a) achieve a high level of control and provide consistent inlet air conditions to the SDCS throughout all tests, and (b) simulate different climates other than the UK; specifically those that favour the use of liquid desiccant based air conditioning systems i.e. high relative humidity. The environmental chamber can create air conditions from 0 to 40 °C and 10–80 % relative humidity. Unlike the IDCS tests, the whole SDCS has been placed in the environmental chamber itself, thus providing a realistic

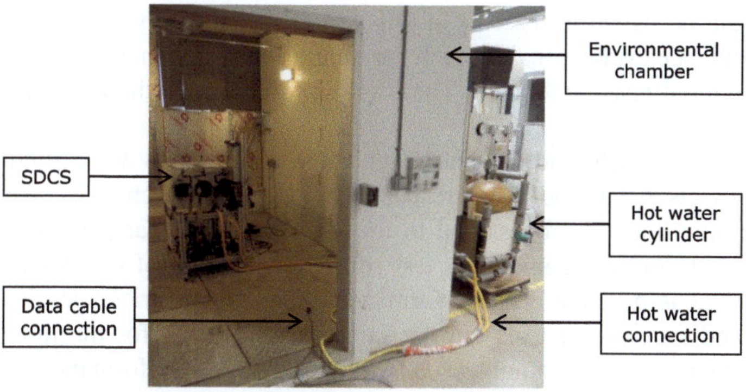

Fig. 6.4 Complete SDCS experimental set-up

representation of the operation of the system under a particular climatic condition. Figure 6.4 provides a photograph of the complete SDCS laboratory experimental set-up, including the environmental chamber and hot water cylinder.

First, the temperature and relative humidity of the environmental chamber are set. Depending on the requirements it can take up to 1 h to achieve stable and homogenous air conditions inside. The temperature and relative humidity within the chamber is shown on the chambers display panel, and is cross checked against the four Vaisalia HMP110 humidity and temperature probes on the SDCS and an RS 1365 handheld humidity-temperature meter. Once the desired air conditions are achieved, and depending on the test variable under investigation, the SDCSs operation is set accordingly and run at that condition.

For the regenerator tests, a vented 120 L hot water cylinder with a 3 kW electrical immersion heater is used as the thermal input to the SDCS. This is placed outside of the chamber and piped though the door to the SDCS using a flexible PVC hot water hose. Before the start of a regenerator test the hot water tank heater and circulation pump (H/P) are switched on. A by-pass loop is used to circulate the water around the tank until it reaches the desired temperature for the particular test. A control valve (V4) is then used to provide the desired hot water flow to the SDCS. The tank thermostat is set according to the required flow temperature, and is checked at regular intervals. Additional control of the flow temperature is achieved by turning the heater on and off manually.

The desiccant solution concentration in both the dehumidifier and regenerator tanks are recorded at the start, mid-point and end of each test, and the results recorded. The air velocity is measured at each duct outlet and recorded at the beginning of a test, and the result recorded. Multiplication of the average air velocity by the air duct area provides the volumetric air flow through the cores. The desiccant solution and water volumetric flow are measured at the start of a test, and the flow indictors checked periodically throughout a test. Depending on the test variable being investigated, tests last for around 30–60 min or until steady-state dehumidifier outlet conditions are achieved for extended periods (30 min or more). Data is logged every 10 s in this period. Only steady state-data is used in the performance evaluation. For each variable investigated there are a minimum of three individual tests conducted. The results presented are the average of each of these tests.

Testing of the dehumidifier and regenerator components are carried out separately in Sects. 6.3.1 and 6.3.2 respectively. This approach was taken to simplify the experimental process and to focus on the particular performance of the component. Using the component test results, evaluation of the complete SDCS in Sect. 6.3.3 is possible. The performance evaluation metrics used in the dehumidifier and regenerator tests are the same as those used for the modelling and IDCS evaluation. The equations used to determine these performance metrics may be referred to in Sect. 3.3.3. In the dehumidifier moisture removal rate, change in absolute humidity of air, latent effectiveness and dehumidifier cooling output are evaluated. In the regenerator: moisture addition rate, latent effectiveness, regenerator thermal input and water temperature difference across the plate heat exchanger are evaluated. For the overall SDCS performance, the thermal and electrical COP (COP_{th}) and (COP_{el}) are used, and are previously defined in Eqs. 4.30 and 4.31 respectively. The thermophysical properties of the humid air are determined from in-built functions in EES (see Appendix 2 for reference), and the thermophysical properties of the desiccant solution are determined from linear regression curve fits to published data (James 1998; Melinder 2007), presented in Sect. 3.2 and Appendix 1. Uncertainty analysis is used to provide a measure of the experimental error associated with the calculated values, using the propagation of error formula, provided in Eq. 5.3. The maximum relative uncertainty values for the dehumidifier and regenerator performance studies are presented in their respective experimental results section.

Section 6.2.2 has described the experimental method used to test the SDCS. Next, Sect. 6.3 presents the results and analysis from the testing of the SDCS.

6.3 SDCS Results and Analysis

Testing of the SDCS is carried out within the environmental chamber. The dehumidifier and regenerator component investigations are conducted separately, presented in Sects. 6.3.1 and 6.3.2 respectively. This approach was taken to simplify the experimental process and to focus on the particular performance of the component. Once the performance of the dehumidifier and regenerator are tested and their performance understood, such that the two processes may be balanced, evaluation of the complete SDCS in Sect. 6.3.3 is possible. Section 6.3.4 discusses a novel liquid desiccant tri-generation system operating scenario, with particular reference to the advantages of integrating SOFC and liquid desiccant air conditioning technology in a building context.

Throughout the testing process, consideration is made to tri-generation system integration. This primarily focusses on the required desiccant solution volumetric flow, and the regenerator thermal input values (hot water flow temperature and volumetric flow). Furthermore, dehumidifier and regenerator simulation results have been provided to validate the developed models presented in Chap. 3, operating with a potassium formate solution. The simulations use the average inlet air temperature, relative humidity and solution temperature/mass concentration. During evaluation of the SDCS the recorded desiccant solution mass concentration ranged from 0.65 to 0.7. A challenging aspect of the experimental work was measuring and maintaining a similar desiccant solution mass concentration between tests. As a result, some variation in experimental results between tests has been observed, however these have been accounted for in the average results presented.

Next, Sect. 6.3.1 presents the results and analysis from the dehumidifier component investigation.

6.3.1 SDCS Dehumidifier Component Analysis

This section presents the results and analysis from the dehumidifier component investigation. The analysis investigates the impact changes in environmental and operating conditions have on the performance of the dehumidifier. The variables considered are inlet air temperature, inlet air relative humidity, air volumetric flow and desiccant solution volumetric flow. Unless otherwise varied, Table 6.1 lists the constants used in the dehumidifier investigation.

The key dehumidifier performance metrics considered in this section are moisture removal rate, change in absolute humidity of air across the dehumidifier, latent (dehumidifier) effectiveness, dehumidifier cooling output and COP_{el}. Figure 6.5a presents the performance of the dehumidifier over a one hour period, with an average inlet temperature of 31.01 °C and relative humidity of 69 %. The average moisture removal rate and latent effectiveness over this period was 0.2106 g s^{-1} and 32.08 % respectively. The average cooling output during this period was 527 W. The data is provided to illustrate the steady-state operation of the dehumidifier. Figure 6.5b demonstrates the dehumidification process on a psychrometric chart.

Table 6.1 Dehumidifier operational constants

Variable	Value
Inlet air temperature (°C)	30
Inlet air relative humidity (%)	70
Air volumetric flow (m^3 h^{-1})	256
Desiccant solution volumetric flow (L min^{-1})	3.2
Desiccant solution mass concentration	0.65–0.7
Evaporative cooler air volumetric flow (m^3 h^{-1})	245
Evaporative cooler volumetric water flow (L min^{-1})	1

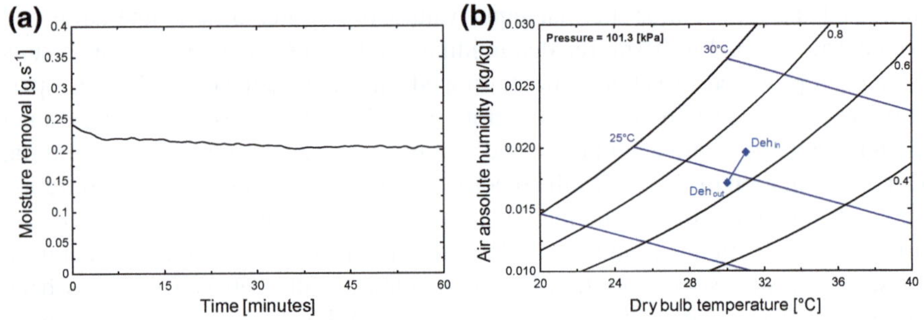

Fig. 6.5 Steady state performance of the SDCS dehumidifier over a 1 h period **a** moisture removal rate and, **b** dehumidifier psychrometric process

6.3.1.1 SDCS Dehumidifier Inlet Air Condition Effect

Figure 6.6 presents the effect of inlet air temperature and relative humidity on dehumidifier performance. The temperature and relative humidity range studied was 25–35 °C and 60–80 % relative humidity respectively. Figure 6.6a shows moisture removal rate increases as both the inlet air temperature and relative humidity increases. Over the temperature range 25–35 °C and at constant RH of 60 %, the moisture removal rate increased from 0.1655 to 0.2258 g s^{-1}. At a

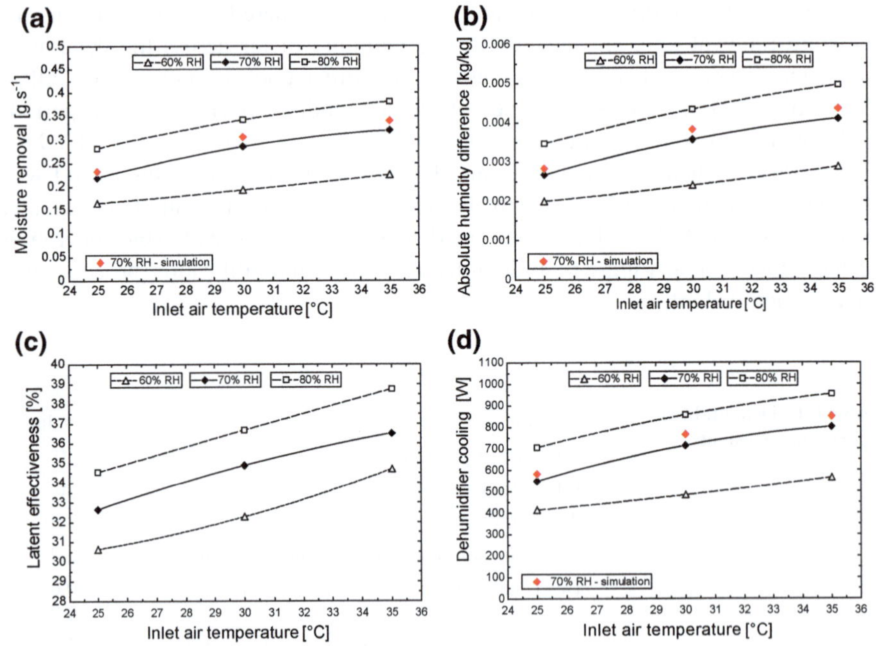

Fig. 6.6 SDCS dehumidifier performance with inlet air conditions

constant RH of 70 % moisture removal rate increased from 0.2195 to 0.3203 g s^{-1} and at a constant RH of 80 % moisture removal rate increases from 0.2823 to 0.3813 g s^{-1}. Similarly, Fig. 6.6b shows that the change in absolute humidity of the air across the dehumidifier increases as the inlet air temperature and relative humidity increases. For a constant RH of 60 % the change in absolute humidity of the air increases from 0.002007 to 0.002876 kg/kg, for a constant RH of 70 % the change in absolute humidity increases from 0.00268 to 0.004095 kg/kg and for a constant RH of 80 % the change in absolute humidity increases 0.003474 to 0.004953 kg/kg. Figure 6.6c shows that latent (dehumidifier) effectiveness increases as both the inlet air temperature and relative humidity increases. For a constant RH of 60 %, latent effectiveness increases from 30.64 to 34.7 %, for a constant RH of 70 % latent effectiveness increases from 32.66 to 36.51 % and for a constant RH of 80 % latent effectiveness increases from 34.55 to 38.75 %.

Because the entire SDCS is placed within the environmental chamber, the water for the evaporative cooler and desiccant solution for the dehumidifier are at ambient temperature i.e. the temperature inside the chamber; this had an impact on the moisture absorption capacity of the desiccant solution and the potential for sensible cooling (little or no sensible cooling was achieved). Therefore, the cooling output reported for the dehumidifier is focussed on the latent cooling potential. When the entire SDCS is placed within the chamber, the evaporative cooler only provides around 80–150 W of cooling over the studied range. This is because the direct evaporative cooling process is not as effective under humid climatic conditions, due to a higher wet-bulb temperature. As a result, the evaporative cooling provided is not enough to produce a sufficient solution temperature decrease and to provide sensible cooling to the supply air in the dehumidifier. Figure 6.6d shows that the dehumidifier cooling output increases as both the inlet air temperature and relative humidity increases. Over the temperature range 25–35 °C and at a constant RH of 60 % the dehumidifier cooling output increases from 414 to 565 W. At a constant RH of 70 % the dehumidifier cooling output increases from 550 to 801 W and at a constant RH of 80 % the dehumidifier cooling output increases from 706 to 954 W. The COP_{el} of the dehumidifier increases with increasing inlet air temperature and relative humidity, from 3.8 at 25 °C 60 % RH to 8.7 at 35 °C 80 % RH. At the conditions investigated it is concluded that the dehumidifier performance increases with increasing temperature and relative humidity.

The partial pressure of water vapour in air increases as the inlet air temperature and relative humidity increases, leading to a greater vapour pressure differential between the air and desiccant solution. An increase in the vapour pressure differential between the air and desiccant solution leads to increased mass transfer and thus a greater moisture removal rate and change in absolute humidity of the air across the dehumidifier. The desiccant system is therefore well suited to hotter more humid climates such as China or costal Mediterranean regions. However, an increase in dehumidification capacity will result in a greater dilution of the desiccant solution. For successful tri-generation system integration, consideration needs to be given to whether the regenerator moisture addition rate achievable with the

SOFC CHP thermal output can match the mass removal rate in the dehumidifier. If it cannot, the operation of the dehumidifiers needs to be limited in order to achieve balanced continuous operation.

6.3.1.2 SDCS Dehumidifier Volumetric Air Flow Effect

Figure 6.7 shows how changes in volumetric air flow effect dehumidifier performance. The volumetric air flow range investigated was 102–256 $m^3\,h^{-1}$. Figure 6.7a shows that moisture removal rate increases as the volumetric air flow increases. Moisture removal rate increased from 0.1114 $g\,s^{-1}$ at 102 $m^3\,h^{-1}$ to a maximum of 0.2859 $g\,s^{-1}$ at 256 $m^3\,h^{-1}$. However, Fig. 6.7b shows that as the air volumetric air flow through the dehumidifier increases, the change in absolute humidity of the air across the dehumidifier decreases from 0.003511 kg/kg at 102 $m^3\,h^{-1}$ to 0.003102 kg/kg at 256 $m^3\,h^{-1}$ i.e. the outlet air absolute humidity is higher. This is because the mass of air passing through the dehumidifier is increasing and thus the capacity of the dehumidifier to reduce the air absolute humidity decreases.

Figure 6.7c shows that as the volumetric air flow increases, the latent (dehumidifier) effectiveness decreases from 42.48 % at 102 $m^3\,h^{-1}$ to 34.98 % at 256 $m^3\,h^{-1}$. Figure 6.7d shows that the dehumidifier cooling output increases as the volumetric air flow increases from 278.6 W at 102 $m^3\,h^{-1}$ to 714.9 W at

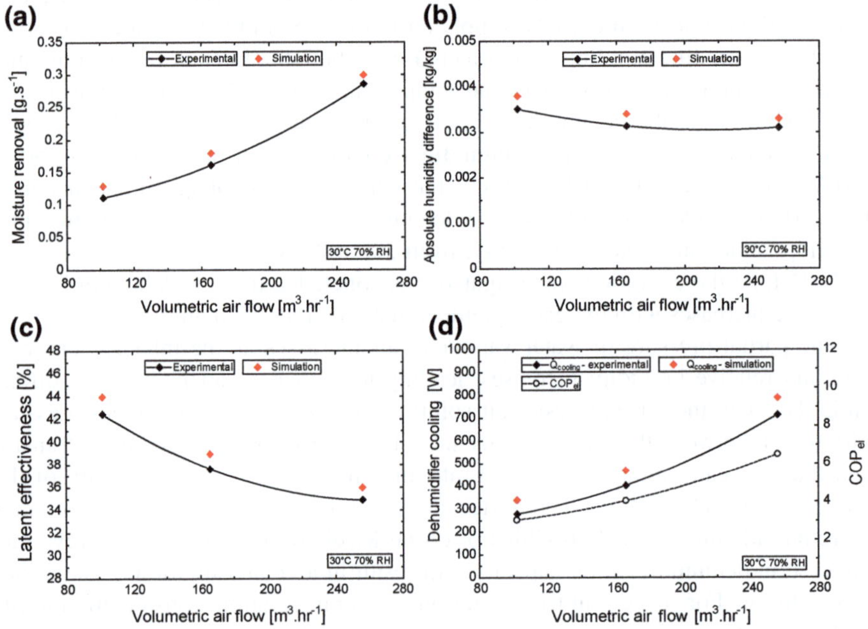

Fig. 6.7 SDCS dehumidifier performance with inlet air volumetric flow

256 m^3 h^{-1}. The observed cooling output increases with the volumetric air flow, even though the outlet air absolute humidity is higher, because there is a greater mass of air being conditioned. Figure 6.7d also shows the COP$_{el}$ of the dehumidifier. The COP$_{el}$ increases with increasing volumetric air flow, from 3.03 at 102 m^3 h^{-1} to 6.49 at 256 m^3 h^{-1}. However, the increase in COP$_{el}$ is not proportional to the increase in dehumidifier cooling output. This is because the electrical load increases with increasing air volumetric flow, from 92 W at 102 m^3 h^{-1} to 110 W at 256 m^3 h^{-1}.

In the dehumidifier, the selection of a suitable volumetric air flow is dependent upon the desired outlet air condition and the volume of air required for the particular application. However, for maximising the cooling achieved, and thus maximising efficiency in a tri-generation system application, the volumetric air flow should be as large as possible.

6.3.1.3 SDCS Dehumidifier Desiccant Solution Volumetric Flow Effect

Figure 6.8 presents the effect of desiccant volumetric flow on dehumidifier performance. The volumetric desiccant flow range studied was 1.5– 3.5 L min^{-1}. Figure 6.8a shows that as the desiccant volumetric flow was increased from 1.5 to 3.5 L min^{-1}, the moisture removal rate increased from 0.2489 to 0.3112 g s^{-1}. Similarly, Fig. 6.8b shows that as the desiccant volumetric flow increased from

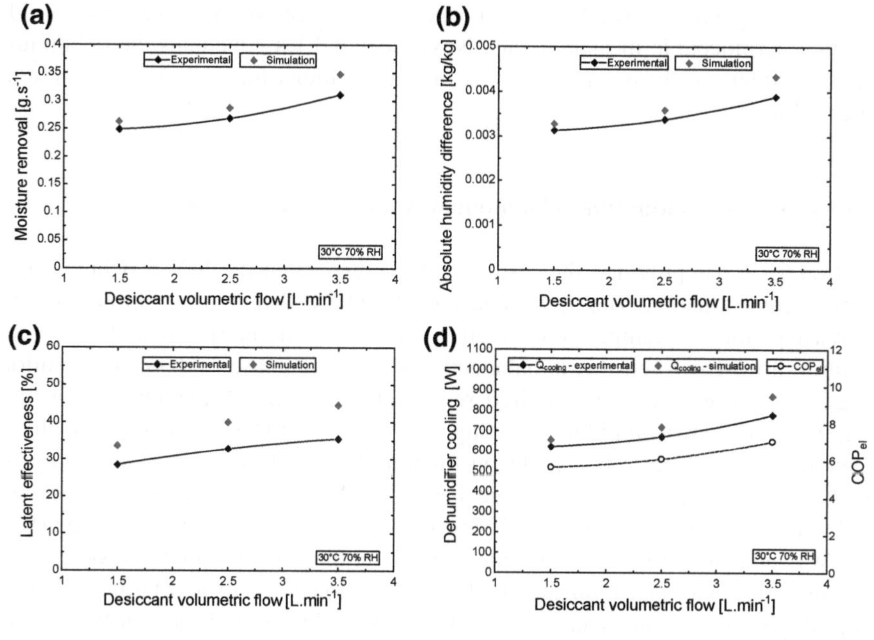

Fig. 6.8 SDCS dehumidifier performance with desiccant volumetric flow

1.5 to 3.5 L min^{-1}, the absolute humidity difference of the air passing through the dehumidifier increased from 0.003132 to 0.003887 kg/kg.

Figure 6.8c shows that the latent (dehumidifier) effectiveness increased as the desiccant solution volumetric flow increased, from 28.42 % at 1.5 L min^{-1} to 35.51 % at 3.5 L min^{-1}. Figure 6.8d shows that the dehumidifier cooling output increases with desiccant solution volumetric flow, from 622.5 W at 1.5 L min^{-1} to 778.2 W at 3.5 L min^{-1}. Figure 6.8d also shows the COP_{el} of the dehumidifier increasing with desiccant solution volumetric flow, from 5.66 at 1.5 L min^{-1} to 7.07 at 3.5 L min^{-1}. As the desiccant solution volumetric flow increases, the solution's capacity for moisture absorption increases. As a result elevated dehumidifier performance is observed with increased desiccant solution volumetric flow.

Figure 6.8 shows the dehumidifier performance increasing with desiccant solution volumetric flow. However, the desiccant solution volumetric flow cannot be increased too far due to issues of flooding of the membrane HMX core and thus desiccant entrainment into the supply air stream; an undesirable and wasteful situation. Furthermore, in a tri-generation system application, to ensure balanced system operation the desiccant solution volumetric flow in the dehumidifier will be dictated by the thermal energy available from the prime mover technology (SOFC). Figure 6.12a shows the impact of changing the regenerator desiccant solution volumetric flow in terms of the moisture addition rate; this needs to be balanced with the dehumidifier. Figure 6.12b shows the thermal energy input required; this needs to match the SOFC thermal output.

Across all dehumidifier tests, the maximum calculated relative uncertainties in the dehumidifier MRR, η_L, $\dot{Q}_{cooling}$ and COP_{el} are ±15.4, ±20, ±15.5 and ±28.5 % respectively. The largest sources of error come from the relative humidity and thermocouple measurements. Next, Sect. 6.3.1.4 concludes the dehumidifier component analysis and discusses the implications for tri-generation system integration.

6.3.1.4 SDCS Dehumidifier Component Analysis Conclusions

Section 6.3.1 has provided the results and analysis from the SDCS dehumidifier component evaluation. The performance of the SDCS dehumidifier improves with increasing inlet air temperature, relative humidity, volumetric air and desiccant solution flow. However, this is balanced against an increase in desiccant solution dilution and thus a greater required regenerator capacity. A maximum moisture removal rate and dehumidifier cooling output of 0.3203 g s^{-1} and 954 W have been achieved respectively. The latent (dehumidifier) effectiveness is in the range of 30–40 %, typical of a membrane based HMX. Because the SDCS is tested within the environmental chamber, very little evaporative cooling is achieved (80–150 W), and thus little or no air side sensible cooling is provided. However, this is a realistic result and replicates real life operation in a hot and humid climate. The SDCS has demonstrated high COP_{el} values of up to 8.7, demonstrating the potential of the SDCS in the provision of efficient and environmentally benign air conditioning, particularly in high humidity applications.

Dehumidifier performance will need to be balanced by the available regeneration capacity obtainable from the SOFC CHP system during tri-generation system integration. Next, Sect. 6.3.2 presents the results and analysis from the regenerator component investigation.

6.3.2 SDCS Regenerator Component Analysis

This section presents the results and analysis from the regenerator component investigation. The analysis investigates the impact changes in environmental and operating conditions have on the performance of the regenerator. The variables considered are inlet air temperature, inlet air relative humidity, air volumetric flow and desiccant solution volumetric flow. Two further tests, aiming to consider the impact of the SOFC CHP system operation on regenerator performance are also conducted. These are hot water flow temperature and hot water volumetric flow in the heating circuit. The regenerator variables investigated have been selected because they impact regenerator performance, but are also dictated by the SOFC CHP system performance. Optimisation of both sets of variables will facilitate effective pairing of the two systems during tri-generation system integration. Unless otherwise varied, Table 6.2 lists the constants used in the regenerator investigation.

The base condition for the inlet air to the regenerator is that of ‘room’ air. As discussed in Chaps. 3 and 4, a lower temperature and relative humidity will improve regenerator performance. During the regenerator tests, the hot water flow temperature and hot water volumetric flow range used is 45–60 °C and 0.5–3.5 L min^{-1} respectively. The thermal input values used in the regenerator evaluation are based upon those typical of a SOFC CHP system found in the literature, such as the BlueGEN unit presented in Chap. 7. This assumption aims to simulate the SDCS operation in a SOFC tri-generation system application, and thus facilitate successful tri-generation system integration in Chap. 7.

The key regenerator performance metrics considered in this section are: moisture addition rate, latent (regenerator) effectiveness and regenerator thermal input. The regenerator thermal input has been measured on the hot water heating side

Table 6.2 Regenerator operational constants

Variable	Value
Inlet air temperature (°C)	26
Inlet air relative humidity (%)	60
Air volumetric flow (m^3 h^{-1})	256
Desiccant solution volumetric flow (L min^{-1})	3.2
Desiccant solution mass concentration (–)	0.65–0.7
Regenerator water volumetric flow (L min^{-1})	2
Regenerator water flow temperature (°C)	55

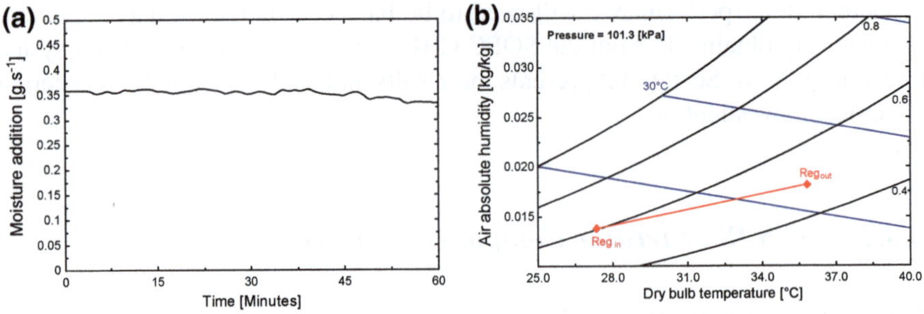

Fig. 6.9 Steady state performance of the SDCS regenerator over a 1 h period **a** moisture addition rate and, **b** regenerator psychrometric process

using Eq. 5.4. Where appropriate, the change in water temperature across the regenerator plate heat exchanger (PX2) is also included (ΔT_{water}). This is to assist with optimisation of the regeneration process with the SOFC CHP system's thermal output during tri-generation system integration.

Figure 6.9a presents the performance of the regenerator over a 1 h period, with an average inlet temperature of 27.36 °C and relative humidity of 60.57 %. The average moisture addition rate and latent (regenerator) effectiveness over this period was 0.3527 g s^{-1} and 34.7 % respectively. The average regenerator thermal input during this period was 1343 W. Figure 6.9b demonstrates the regeneration process on a psychrometric chart. The data is provided to illustrate the steady-state operation of the regenerator during laboratory experimentation. Throughout all tests the regenerator plate heat exchanger (PX2) effectiveness ranged from 60 to 80 %.

6.3.2.1 SDCS Regenerator Inlet Air Condition Effect

Figure 6.10 presents the effect of inlet air relative humidity on regenerator performance. The relative humidity range studied was 50–80 % at a constant temperature of 30 °C. Figure 6.10a shows that as the inlet air relative humidity

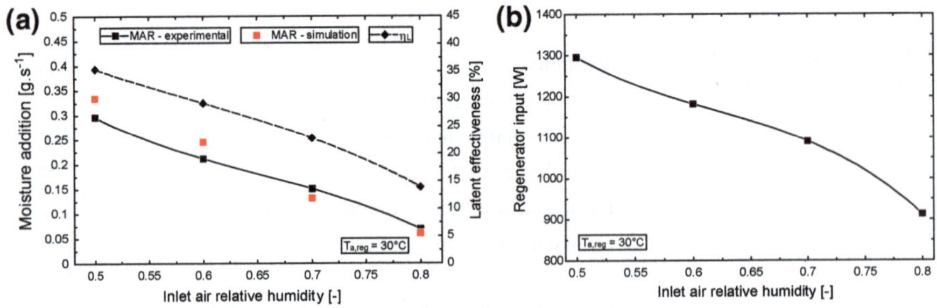

Fig. 6.10 SDCS regenerator performance with inlet air relative humidity

increases, the moisture addition rate decreases from 0.2968 g s^{-1} at 50 % RH to 0.06915 g s^{-1} at 80 % RH. Figure 6.10a also shows that as the inlet air RH increases the latent (regenerator) effectiveness decreases from 35.43 % at 50 % RH to 13.9 % at 80 % RH. Figure 6.10b shows that as the inlet air RH increases, the regenerator thermal input decreases, from 1296 W at 50 % RH to 912.9 W at 80 % RH. The temperature difference of the heating water flowing through the regenerator plate heat exchanger (PX2) ranged from 6.5 to 9.3 °C, an important consideration for tri-generation system integration.

It is evident that performance of the regenerator is greater at lower inlet air RH. Furthermore, if we compare results at air temperature of 30 °C and RH of 60 % (Fig. 6.10a) with results at an air temperature of 26 °C and RH of 60 % (Fig. 6.11a) at 256 $m^3 h^{-1}$ it is apparent that the regenerator performance is greater with a lower inlet air temperature. This is because a lower inlet air temperature and relative humidity will exhibit a lower air side vapour pressure. Thus, at a constant solution inlet condition to the regenerator, a greater vapour pressure differential will exist, and therefore increased potential for mass transfer from the desiccant solution to the air stream. Therefore in a tri-generation system application, it is beneficial to system performance to operate the regenerator on the lowest air temperature and relative humidity possible. This will mean for a given thermal output from the SOFC CHP system, the moisture addition rate in the regenerator will be higher, and thus the allowable moisture removal rate in the dehumidifier can be higher. This will lead to a greater cooling output for the given thermal input, and thus elevated tri-generation system efficiency. Although dehumidifier cooling output increases in hot and humid conditions, when operating the regenerator on fresh outside air regeneration capacity reduces with increasing inlet air relative humidity and temperature and thus liquid desiccant system performance will be poorer in hot and high humid climates.

6.3.2.2 SDCS Regenerator Volumetric Air Flow Effect

Figure 6.11 presents the effect volumetric air flow has on regenerator performance. The volumetric air flow range studied is 102–256 $m^3 h^{-1}$. Figure 6.11a shows

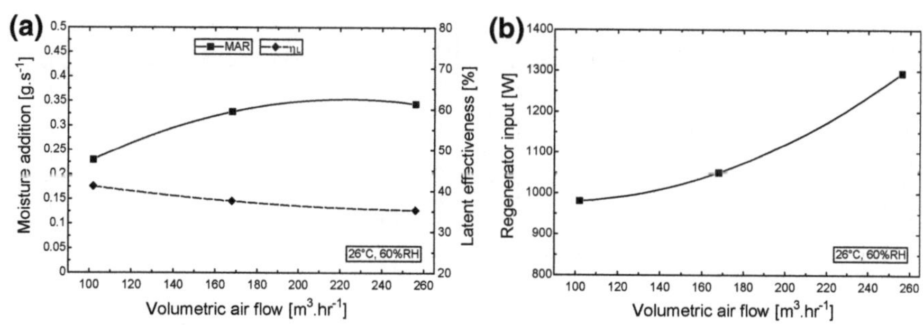

Fig. 6.11 SDCS regenerator performance with inlet air volumetric flow

as the inlet volumetric air flow increases, the moisture addition rate increases slightly from 0.2299 g s^{-1} at 102 m^3 h^{-1} to 0.3442 g s^{-1} at 256 m^3 h^{-1}. However, Fig. 6.11a shows that as the inlet air volumetric flow increases, the latent (regenerator) effectiveness decreases. This is because latent effectiveness is calculated with respect to the absolute humidity difference of the regenerator airstream. As the volumetric air flow is increased the absolute humidity difference (increase) of the regenerator airstream actually reduces, and thus the latent effectiveness is reduced. Figure 6.11b shows that as the air volumetric flow increases, the regenerator thermal input also increases from 980.6 to 1294 W. The temperature difference of the heating water flowing through the regenerator plate heat exchanger (PX2) ranges from 7 to 9 °C.

The regeneration process is used to re-concentrate the desiccant solution, and thus allow continuous operation of the dehumidifier. Therefore the regenerator outlet air condition is not of importance, however the moisture removed from the desiccant solution is. As a result, the regenerator should be operated at the highest volumetric air flow, to facilitate increased regeneration. However, during tri-generation system integration, this will need to be balanced with the available thermal input from the SOFC CHP system.

6.3.2.3 SDCS Regenerator Desiccant Solution Volumetric Flow Effect

Figure 6.12 shows the effect of volumetric desiccant flow on regenerator performance. The volumetric desiccant flow range studied was 1.5–3.5 L min^{-1}. Figure 6.12a shows that as the volumetric desiccant flow is increased, moisture addition rate and latent effectiveness increases from 0.08311 to 0.3442 g s^{-1} and 15.73 to 39.46 %, respectively. Figure 6.12b shows that as the volumetric desiccant flow increases, regenerator thermal input and the water temperature difference increases from 708.3 to 1294 W and 5.08 to 9.27 °C respectively. As the desiccant solution volumetric flow increases, the solution capacity for moisture de-sorption increases. As a result elevated regenerator performance is observed with increased desiccant solution volumetric flow. However, this is balanced against an increase

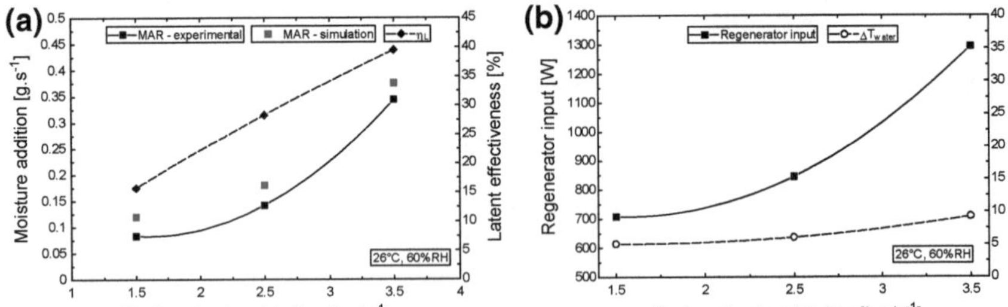

Fig. 6.12 SDCS regenerator performance with desiccant volumetric flow

in the regenerator thermal input as the volumetric flow increases and a higher water temperature difference across the regenerator plate heat exchanger.

Selection of a suitable desiccant solution volumetric flow in the regenerator, particularly for effective tri-generation system integration, is dependent upon achieving sufficient wetting of the membrane HMX, balancing of the moisture addition with the moisture removal in the dehumidifier, whilst matching the regenerator thermal requirement and water temperature difference with the SOFC CHP systems thermal input.

6.3.2.4 SDCS Regenerator Water Flow Temperature Effect

Figure 6.13 shows the impact of regenerator water flow temperature on regenerator performance. The water flow temperature range investigated was 45–60 °C. This range was selected because it covers a typical temperature output of a SOFC WHR circuit. Figure 6.13a shows that as the inlet water temperature is increased, moisture addition rate increased from 0.2156 g s^{-1} at 45 °C up to 0.4331 g s^{-1} at 60 °C. Similarly, the latent (regenerator) effectiveness increased with inlet water temperature, from 32.83 % at 45 °C to 36.52 % at 60 °C.

Over the range investigated, the temperature of the desiccant solution entering the regenerator HMX core was 40–53 °C. An increase in the regenerator water flow temperature means the desiccant solution is raised to a higher temperature. An increase in desiccant solution temperature will result in a higher solution side vapour pressure. At a constant inlet air condition, this will lead to a greater vapour pressure differential between the solution and air, and thus an increase in mass transfer. Figure 6.13b shows that regenerator thermal input increases as the inlet water temperature increases, from 958 W at 45 °C to 1588 W at 60 °C. The difference in water temperature across the regenerator plate heat exchanger (PX 2) increases from 6.7 to 11.38 °C.

During tri-generation system integration, the regenerator water flow temperature will be dependent on the SOFC CHP system itself; the temperature may be

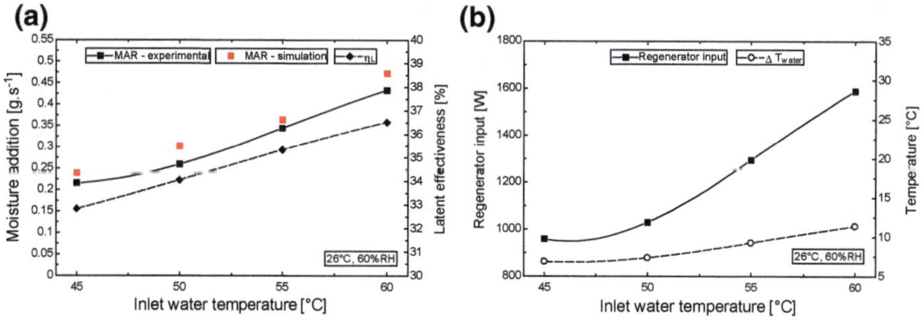

Fig. 6.13 SDCS regenerator performance with water flow temperature

variable to some degree through the volumetric flow in the heat recovery circuit. As a result, the other variables investigated in this section will need to be selected appropriately in order to enable balanced operation. Regeneration of the desiccant solution is possible with a hot water inlet temperature as low as 45 °C, demonstrating the potential for tri-generation system integration.

6.3.2.5 SDCS Regenerator Heat Exchanger Volumetric Flow Effect

Figure 6.14 shows the impact of regenerator heat exchanger water volumetric flow on regenerator performance. The volumetric flow range investigated was 0.5–3.5 L min^{-1}. This range was selected because it covers a typical volumetric flow of a SOFC WHR circuit. Figure 6.14a shows that as the water volumetric flow increases from 0.5 to 3.5 L min^{-1}, both the moisture addition rate and latent (regenerator) effectiveness increase from 0.06061 to 0.3565 g s^{-1} and 30.17 to 37.97 % respectively. Over the range investigated, the temperature of the desiccant solution entering the regenerator HMX core was 41–51 °C. An increase in desiccant solution temperature results in an increase in mass transfer from the solution to the air.

Figure 6.14b demonstrates that as water volumetric flow is increased, the regenerator thermal input increases from 659.2 to 1749 W, however the water temperature difference across the regenerator plate heat exchanger (PX 2) decreases from 18.88 to 7.16 °C. The selection of a suitable regenerator heat exchanger water volumetric flow will be dependent upon achieving a hot water flow temperature from the SOFC CHP system that will facilitate effective desiccant solution regeneration. Testing of the SOFC CHP system is presented in Chap. 7.

Across all regenerator tests, the maximum calculated relative uncertainties in the regenerator MAR and η_L are ±7.8 and ± 22.8 % respectively. Next, Sect. 6.3.2.6 concludes the regenerator component analysis and discusses the implications for tri-generation system integration.

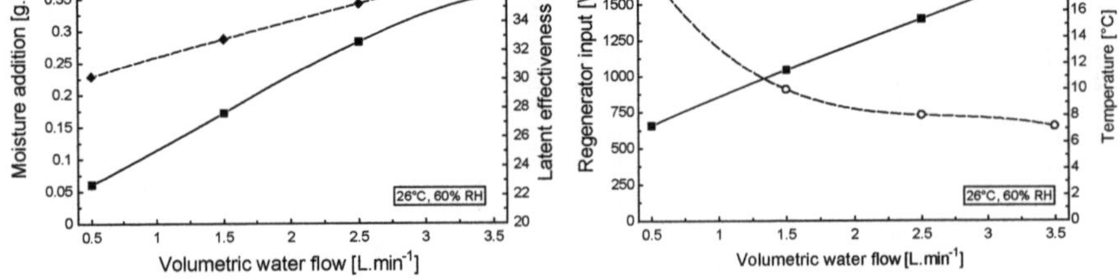

Fig. 6.14 SDCS regenerator performance with heat exchanger volumetric water flow

6.3.2.6 SDCS Regenerator Component Analysis Conclusions

Section 6.3.2 has provided the results and analysis from the SDCS regenerator component evaluation. The results and analysis show regeneration of a potassium formate solution at a 0.65–0.7 solution mass concentration is possible with a hot water flow temperature as low as 45 °C and improves as the hot water flow temperature increases. Regenerator performance improves with a lower inlet air temperature and relative humidity. As a result, the work in this thesis considers extract room air as the regenerator inlet air stream for tri-generation system integration. Regenerator capacity is greater with increasing volumetric air and desiccant solution flow. However, increasing these variables demonstrates an increase in the required regenerator thermal input. For effective tri-generation system integration this needs to be balanced against the available thermal input from the SOFC CHP system. Furthermore, regenerator capacity is greater with increasing hot water flow temperature and water volumetric flow in the heating circuit. A maximum moisture addition rate of 0.4331 g s^{-1} has been achieved, with a regenerator thermal input of 1588 W. The latent (regenerator) effectiveness is in the range of 15–40 %. During tri-generation system integration, the performance of the regenerator will be largely dictated by the available thermal energy from the SOFC CHP system. To ensure balanced SDCS operation, this will inform the permissible moisture removal rate that the dehumidifier should be operated at.

Next, Sect. 6.3.3 presents the results and analysis for the complete SDCS operating in a balanced manner.

6.3.3 Complete SDCS Performance Analysis

The aim of this section is to analyse the performance of the dehumidifier and regenerator operating together in the complete SDCS. In order to accurately evaluate the complete SDCS, the moisture removal rate in the dehumidifier must balance the moisture addition rate in the regenerator, thus facilitating continuous operation of the SDCS. Once balanced operation of the SDCS has been achieved, calculation of the required thermal input and COP_{th} is possible. Furthermore, specific consideration is given to the SDCS operating with the SOFC CHP systems thermal output.

The experimental dehumidifier moisture removal rates have been summarised based on inlet air temperature, inlet air relative humidity and desiccant solution volumetric flow. Regenerator moisture addition rates that balance these dehumidifier moisture removal rates have been attained through varying three key regenerator variables (1) regenerator desiccant solution volumetric flow, (2) hot water volumetric flow in the heating circuit and (3) hot water flow temperature. Balancing of the SDCS by means of these three key regenerator variables allows identification of the optimum means in which to operate the regenerator under a particular environmental or operational condition. The objective is to minimise the required thermal input for a particular cooling output, and thus maximise the COP_{th}. In a tri-generation system, this will facilitate improved overall efficiency. Furthermore, this approach to system analysis will illustrate whether the SDCS will perform effectively with the SOFC CHP system's thermal output once integrated into a tri-generation system in Chap. 7. As previously explained in Sect. 6.2.2, the entire SDCS tests have been conducted within the environmental chamber. As a result, the COP_{th} values presented are realistic, providing an indication of the actual performance of the SDCS under real life environmental conditions. Unless otherwise varied, Tables 6.1 and 6.2 list the operational constants for the dehumidifier and regenerator respectively.

Figure 6.15 shows the variation in SDCS performance with dehumidifier inlet air relative humidity at a 30 °C inlet air temperature. The SDCS performance is presented with respect to regenerator desiccant solution volumetric flow (Fig. 6.15a), water volumetric flow in the heating circuit (Fig. 6.15b) and the hot water flow temperature in the heating circuit (Fig. 6.15c). The COP_{th} of the SDCS increases as dehumidifier inlet air relative humidity is increased. Figure 6.15a shows that as the dehumidifier inlet air RH is increased, the required regenerator desiccant solution volumetric flow also increases. At a dehumidifier RH of 60 %, the regenerator solution flow required was found to be 2.834 L min^{-1} (less than the dehumidifier), at 70 % RH the required regenerator solution flow was found to be 3.274 L min^{-1} (roughly equal to the dehumidifier) and at 80 % RH the

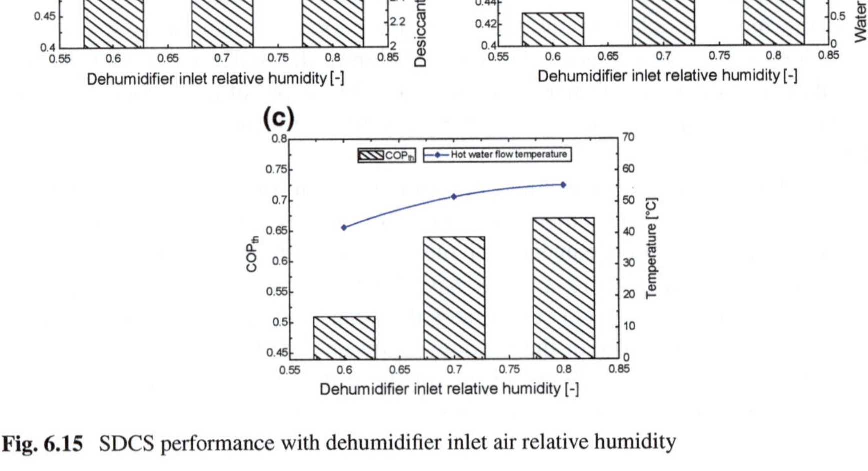

Fig. 6.15 SDCS performance with dehumidifier inlet air relative humidity

required regenerator solution flow was found to be 3.495 L min^{-1} (more than the dehumidifier). Figure 6.15b shows that the hot water volumetric flow in the heating circuit needs to be increased as the dehumidifier inlet air relative humidity is increased, from 1.683 L min^{-1} at 60 % RH, to 2.526 L min^{-1} at 70 % RH and to 3.232 L min^{-1} at 80 % RH. Figure 6.15c shows that the hot water flow temperature needs to be increased as the dehumidifier inlet air relative humidity increases, from 41.92 °C at 60 % RH, to 51.58 °C at 70 % RH and to 55.24 °C at 80 % RH. Over the range of dehumidifier inlet air RH values investigated the regenerator thermal input ranged from 941 to 1653 W.

At a set inlet dehumidifier air condition of 30 °C and 70 % RH, the COP_{th} values seen in Fig. 6.15a–c are 0.61, 0.51 and 0.64 respectively. Over the relative humidity range investigated, COP_{el} varies between 4.4 and 7.8. Although not shown in Fig. 6.15 COP_{th} of the SDCS improves with dehumidifier inlet air temperature.

Figure 6.16 shows the variation in SDCS performance with dehumidifier desiccant solution volumetric flow. The SDCS performance is presented with respect to regenerator desiccant solution volumetric flow (Fig. 6.16a), water volumetric flow in the heating circuit (Fig. 6.16b) and the hot water flow temperature in the heating circuit (Fig. 6.16c). The COP_{th} of the SDCS increases as dehumidifier desiccant solution volumetric flow is increased. Figure 6.16a shows that as the dehumidifier desiccant solution flow is increased, so does the required desiccant solution flow in the regenerator. The required change in regenerator desiccant solution flow is not proportional to the change in the dehumidifier desiccant

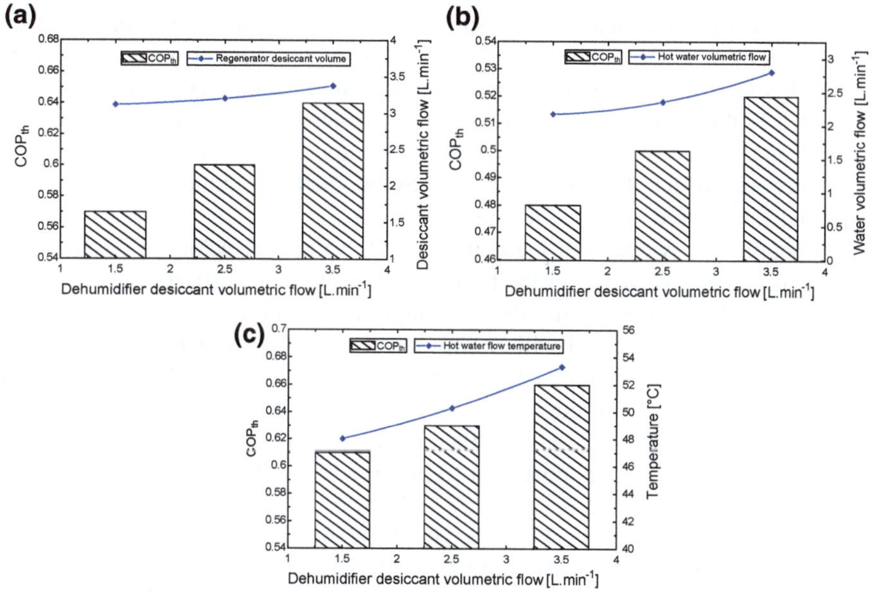

Fig. 6.16 SDCS performance with dehumidifier desiccant solution volumetric flow

solution flow. At a dehumidifier solution flow of 1.5 L min^{-1}, the required regenerator solution flow is 3.114 L min^{-1}, at 2.5 L min^{-1} the required regenerator solution flow is 3.2 L min^{-1} and at 3.5 L min^{-1} the required regenerator solution flow is 3.376 L min^{-1}. Figure 6.16b shows the hot water volumetric flow in the heating circuit needs to be increased as the dehumidifier desiccant solution volumetric flow is increased. At a desiccant flow of 1.5 L min^{-1} the water volumetric flow is 2.17 L min^{-1}. As desiccant volumetric flow increases to 2.5 L min^{-1} the water volumetric flow needs to be increased to 2.353 L min^{-1} and at a desiccant volumetric flow of 3.5 L min^{-1} the water volumetric flow needs to be increased to 2.802 L min^{-1}. Figure 6.16c shows the hot water flow temperature needs to be increased as the dehumidifier desiccant solution volumetric flow is increased, from 48 °C at 1.5 L min^{-1}, 50.27 °C at 2.5 L min^{-1} and to 53.3 °C at 3.5 L min^{-1}. Over the dehumidifier desiccant solution volumetric flow values investigated the regenerator thermal input ranges from 1014 to 1502 W.

At a set inlet dehumidifier desiccant solution volumetric flow of 3.5 L min^{-1}, the COP_{th} values seen in Fig. 6.16a–c are 0.64, 0.52 and 0.66 respectively. Over the dehumidifier desiccant solution volumetric flow range investigated, the COP_{el} varies between 5.7 and 7.1.

The data presented in Figs. 6.15 and 6.16 are representative for the SDCS, illustrating the relationships and trends between the various operational parameters in the SDCS. The results presented demonstrate that all three key regenerator variables used to achieve balanced operation are within an acceptable and realistic operating range whilst adopting practical SOFC thermal input values. This is an essential and crucial conclusion for effective tri-generation system integration. The highest SDCS COP_{th} values are achieved when adjusting the hot water flow temperature from the heat source, and the lowest when adjusting the hot water volumetric flow in the heating circuit. The regenerator thermal input used during the SDCS tests is from the hot water cylinder using operational values similar to that of a SOFC CHP system. Tri-generation system integration will replace the hot water cylinder with the SOFC's thermal output. As a result, the SDCS performance may differ slightly if the actual SOFC thermal output is different to the assumed values used during SDCS testing.

Although it is evident that the highest COP_{th} values are achieved when adjusting the flow temperature from the heat source, in the SOFC tri-generation system this may not be possible as the thermal output temperature will either be fixed or in a very narrow range, and will be dependent on the hot water volumetric flow. During tri-generation system integration it is more likely that balancing of the SDCS will be achieved through the selection of an appropriate desiccant solution volumetric flow, both in the dehumidifier and regenerator and hot water volumetric flow in the heating circuit. Based on trying to achieve the highest SDCS COP_{th}, and thus maximising thermal energy utilisation in the tri-generation system, the outlet water temperature from the SOFC WHR circuit should be maximised (within the range investigated), and the WHR volumetric water flow minimised. This is a conclusion demonstrated in Chap. 7. Chapter 7 details the selection of specific SDCS operational values when operating within the tri-generation system context using a fixed regenerator thermal input.

Section 6.3.3 has confirmed experimentally, that SDCS COP_{th} values in the 0.4–0.66 range are achievable with a low grade thermal input (45–60 °C). COP_{th} values presented demonstrate that the integration of liquid desiccant and SOFC technology is feasible, and the development of the novel tri-generation system achievable. System modifications to the evaporative cooler, and thus greater air side sensible cooling could result in higher COP_{th} values approaching 1.0.

Next, Sect. 6.3.4 discusses a novel tri-generation system operating scenario, with particular reference to the advantages of integrating liquid desiccant air conditioning and SOFC technology in a building context.

6.3.4 A Novel Tri-generation System Operating Scenario

In the proposed tri-generation system, the connection between the SOFC and liquid desiccant components is through the regenerator. Section 6.3.2 demonstrates that regeneration capacity is largely dictated by the available (SOFC) thermal input. As highlighted in Chaps. 2 and 4, SOFC CHP systems in built environment applications rarely operate intermittently due to long-start up times and issues of thermal cycling. As a result, there is always a thermal output, but not always a thermal demand, particularly in climates requiring dehumidification/cooling and not heating.

Utilising the constant SOFC thermal output and operating the dehumidifier and regenerator independently or out of sync of one another has the potential to improve tri-generation system performance. Below, two potential nonsynchronous operating scenarios are discussed. Both require the provision of sufficient desiccant solution storage.

(1) Operating the regenerator at a lower capacity, for a greater period of time over and beyond the dehumidifier operating schedule i.e. throughout the day and some of the night. This approach means that the dehumidifier can be operated at a moisture removal rate that is greater than the regenerator moisture addition rate achievable with the instantaneous thermal input from the SOFC. As a result, increased dehumidification cooling capacity is possible with the available thermal input. This is demonstrated experimentally in Chap. 7.

(2) Operating the dehumidifier and regenerator out of sync of one another i.e. day and night respectively, means that the SOFC thermal output can be used in the day for building hot water demands and at night in the regenerator, thus maximising SOFC thermal utilisation over a 24 h period, leading to elevating system efficiency.

In the proposed operating scenarios the re-concentration of the desiccant solution using the constant SOFC thermal output can be considered a form of thermal energy storage. Unlike thermal storage techniques based on sensible energy, a significant advantage of (chemical) thermal energy storage in the form of strong desiccant solution is that there are minimal losses over time.

The nonsynchronous operating scenarios highlight the significant advantage of the novel SOFC liquid desiccant tri-generation system. Other built environment CHP technologies such as Stirling engines or even PEMFCs operate intermittently and thus do not have the opportunity to provide extended thermal input to the liquid desiccant system. Other heat powered air conditioning technologies such as closed cycle vapour absorption would not offer increased cooling potential with extended periods of thermal input. Furthermore, they would not be able to operate at generator temperatures observed in this investigation. Thus the development of a novel SOFC liquid desiccant tri-generation system shows great potential, particularly in built environment applications.

Next, Sect. 6.4 summarises the chapter conclusions and their implications for tri-generation system integration.

6.4 Chapter Conclusion and Implications for Tri-generation Integration

This chapter has presented the evaluation, based on experimental data, of a SDCS. An environmentally friendly potassium formate working fluid has been employed at a solution mass concentration of 0.65–0.7. The SDCS has been developed in response to the shortcomings of the novel IDCS, documented in Chap. 5, and its lack of potential for effective tri-generation system integration. The SDCS experimental set-up, instrumentation and experimental method have all been provided. The thesis has established that during tri-generation system integration it is the operation of the SDCS that will need to be optimised to facilitate successful pairing of the SOFC and liquid desiccant technology. This is due to limited variation in the SOFC CHP system's operation and thus outputs. This chapter has successfully provided a detailed evaluation of the dehumidifier, regenerator and complete SDCS operational performance so that the central thesis aim of the development of an efficient and effective SOFC liquid desiccant tri-generation system can be accomplished. Furthermore, dehumidifier and regenerator simulation results have been presented to validate the models, presented in Chap. 3, operating with a potassium formate solution. Potential reasons for the discrepancies between experimental and simulations have been previously highlighted in Sect. 3.4.3. Two tri-generation system operating scenarios that use concentrated desiccant solution as a form of thermal energy storage have been introduced. The scenarios demonstrate the significant advantage of the novel SOFC liquid desiccant tri-generation system.

The experimental component evaluation has established that the SDCS demonstrates great potential for the development of an efficient and effective SOFC tri-generation system. This is primarily due to good dehumidification capacity and effective regeneration of a potassium formate solution at a 0.65–0.7 solution mass concentration when using regenerator thermal input values typical of a SOFC CHP WHR circuit.

The SDCS shows potential, compared to the IDCS, with regard to tri-generation system integration for the following reasons:

- Effective instantaneous balancing of the dehumidifier and regenerator across a range of environmental and operational values.
- The key regenerator variables used to balance dehumidifier and regenerator operation are within an acceptable and realistic operating range, essential for effective tri-generation system integration.
- Testing of the complete SDCS within the environmental chamber demonstrates good (non-adjusted) COP_{th} values in the region of 0.4–0.66. System modifications to evaporative cooler could result in a COP_{th} approaching 1.0.
- Greater control of individual variables within the desiccant system i.e. dehumidifier and regenerator desiccant solution flow.
- No issues with desiccant solution leakage.

Although the SDCS has been developed with the aim of integration alongside a SOFC into a complete tri-generation system, the compact nature and effective use of low grade waste heat means the SDCS shows significant potential for integration with other CHP prime mover technologies such as ICE or SE. As highlighted throughout the thesis, no work has been found detailing a SOFC liquid desiccant tri-generation system. Simulations in Chap. 4 have proved the tri-generation system concept theoretically. Using operational thermal input values based upon those typical of a domestic scale SOFC CHP system, the SDCS evaluation validates the concept of integrating SOFC and liquid desiccant air conditioning technology. The work presented in this chapter provides a contribution to knowledge with the testing of a liquid desiccant air conditioning system based on specific SOFC thermal input values.

Next, Chap. 7 presents tri-generation system integration and evaluation.

References

Elmer, T., M. Worall, S. Wu, and S. Riffat. 2016b. Experimental evaluation of a liquid desiccant air conditioning system for tri-generation/waste-heat-driven applications. *International Journal of Low-Carbon Technologies* doi: 10.1093/ijlct/ctw012.

James, S. 1998. A new working fluid 'potassium formate' for use in absorption heat pumps. Masters of Philosophy, The University of Nottingham.

Melinder, A. 2007. Thermophysical properties of aqueous solutions used as secondary working fluids. PhD Thesis, KTH Energy and Environmental Technology.

Chapter 7
Tri-generation System Integration and Evaluation

7.1 Introduction

This chapter evaluates the novel tri-generation system concept that is the focus of this thesis. It seeks to address, or at least narrow, the current research gap with an evaluation based on experimental evidence generated in a laboratory setting. In so doing the evaluation quantifies the performance of the system and offers suggestions for future work in the field.

An extensive review of the literature presented in Chap. 2, has shown that an SOFC liquid desiccant tri-generation system will be the first of its kind. Simulations, presented in Chap. 4, of the novel tri-generation system demonstrate the potential for high tri-generation efficiency and PED reductions.

This chapter evaluates two SOFC liquid desiccant tri-generation systems. First, Sect. 7.2 presents a theoretical integration analysis, based on collected empirical SOFC CHP and liquid desiccant data. Second, an experimental tri-generation system is introduced. Section 7.3 provides a summary of this novel experimental tri-generation system configuration, experimental set-up, instrumentation and experimental method. Section 7.4 presents the results and analysis from both the experimental SOFC CHP system sub component tests, and the complete SOFC liquid desiccant tri-generation system tests. Finally, Sect. 7.5 concludes the chapter, with particular respect to the achievement of the thesis aim.

As highlighted in Chaps. 4 and 6, it is the operation of the desiccant system that will need to be optimised to facilitate successful tri-generation system integration. This is due to limited variation in the SOFC CHP system's operation and thus outputs. Chapter 6 established that the SDCS, compared to the IDCS presented in Chap. 5, demonstrates greater potential for tri-generation system integration. This is due to balanced operation, zero desiccant leakage and greater operational control. As a result, it is the SDCS that is adopted for tri-generation system

T. Elmer, *A Novel SOFC Tri-generation System for Building Applications*,
Springer Theses, DOI 10.1007/978-3-319-46966-9_7

integration. To avoid repetition, the Chap. 6 SDCS experimental set-up, experimental method and results are not repeated but referred to where necessary.

7.2 Theoretical Tri-generation System Integration Study

The original SOFC that was intended for tri-generation system integration and field trial testing in a building application was the BlueGEN CHP unit manufactured by Ceramic Fuel Cells Ltd. (CFCL). However, unit failure has meant that the SOFC is no longer available for tri-generation system integration. Although the BlueGEN SOFC is no longer operational, a theoretical integration analysis, based on collected empirical SOFC and SDCS data is presented. Published journal articles, by the author, on SOFC operation in a real building context and tri-generation system integration can be referred to here; Elmer et al. (2015a, 2016a).

BlueGEN is a commercially available SOFC CHP system designed for small to medium scale building applications. Operating on natural gas (NG), the unit can be power modulated from 500 W_e (25 %) to 2 kW_e (100 %), however it achieves its highest net electrical efficiency of 60 % at a 1.5 kW_e output. As a result, CFCL have optimised the default operation of the unit at 1.5 kW_e to provide the highest electrical efficiency and thus greatest economic benefit to the user. The BlueGEN unit consists of 51 planar type YSZ electrolyte layer sets (each layer consist of 4 cells), and operates at 750 °C. Hydrogen is produced from natural gas by internal steam reforming (endothermic) on the fuel cell anode, utilising the heat of the electrochemical reaction (exothermic) to create a chemical combined cycle. The BlueGEN unit was selected for field trial testing because (1) it is commercially available, (2) it fulfils the technical objectives of the thesis, and (3) is certified for domestic building installations and qualifies for the UK FiT (feed-in-tariff); a tariff paid to the consumer per kW_h of generated electricity. The unit is installed at The University of Nottingham's Creative Energy Homes as shown in Fig. 7.1.

Fig. 7.1 BlueGEN SOFC CHP system installed at The University of Nottingham

The SOFC is connected electrically, in parallel, to the national grid in order to export or import power as required. The SOFC is connected to the natural gas grid. A heating water circuit delivers the generated heat from the SOFC unit directly to the homes 300 L hot water cylinder, which is supplemented by an auxiliary gas boiler. For tri-generation system integration, the intention was to install the SDCS outside of the home in-line between the SOFC and hot water cylinder, as shown in Figs. 7.1 and 7.2. Three way diverter valves direct thermal energy to the regenerator during tri-generation system operation.

Figure 7.3 shows field trial electrical performance data collected from the SOFC, using the online CFCL interface, from 24 March 2014 (point 1) to 12 December 2014 (point 8). This is equivalent to 4865 h of operation (8 months

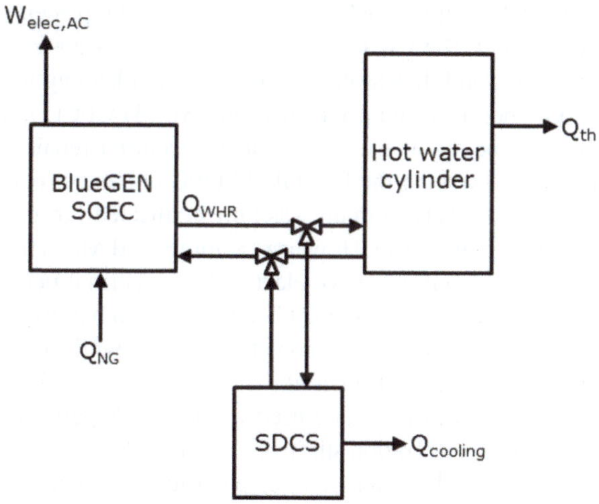

Fig. 7.2 BlueGEN and SDCS connection

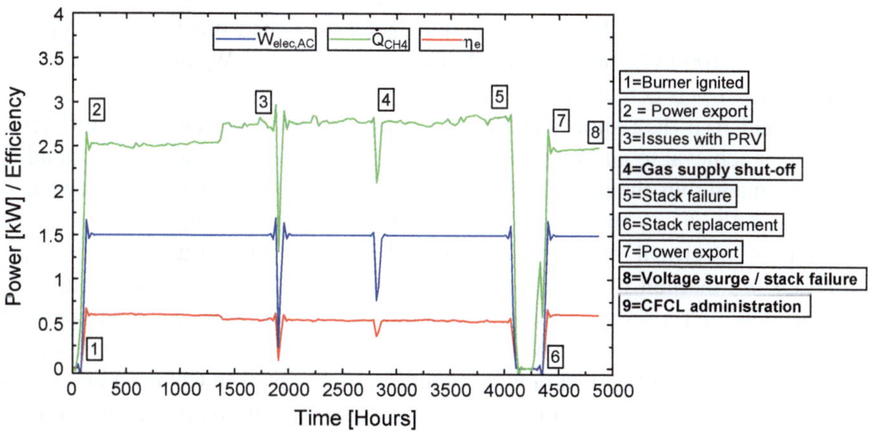

Fig. 7.3 BlueGEN field trial electrical performance data

18 days). During this period the SOFC unit shows stable operation, i.e. electrical efficiency of 55–60 %, with availability for power generation of 91.7 %, therefore demonstrating the potential for the development of an efficient and effective tri-generation system. Due to the time taken to heat the stack to 750 °C and to avoid thermal cycling, the unit operated continuously, always aiming to maintain a 1.5 kW_e output. As seen in Fig. 7.3, as the stack efficiency degrades over time the fuel input is increased to compensate for this. At an electrical efficiency of 60 % the fuel input is 2.5 kW. After 4000 h of operation (point 2–5), the stack displayed an electrical efficiency degradation of approximately 6 %.

Three key events in the lifetime of the unit have meant that it is not available for tri-generation system integration. These events were as follows: (1) an unforeseen gas shut-off causing stack cool down and thermal contraction, leading to an electrical efficiency drop, and eventual stack failure (point 5) and replacement (point 6), (2) A 415 V voltage surge at The Creative Energy Homes causing irrevocable damage to the power electronics and thus stack cool-down, again leading to the requirement of power electronic and stack replacement (point 8). (3) CFCL going into administration, and thus not being able to carry-out the required repair works post voltage surge. At the time of writing the BlueGEN SOFC unit is not operational. During the operational period, the WHR circuit was only connected for a short period, thus there is limited thermal output data. However, Sommer and Messenholler (2013) and Foger (2013) have carried out extensive electrical and thermal performance characterisation of an identical BlueGEN SOFC CHP system in a building application. The results from a BlueGEN power modulation study are shown in Fig. 7.4. During the performance evaluation, Foger (2013) used a 2 L min^{-1} water volumetric flow in the WHR circuit. This is equal to the value used in the SDCS performance analysis in Chap. 6, thus a theoretical integration study is a rational idea.

From Fig. 7.4a it is evident that the net electrical efficiency increases as the electrical capacity increases, from 14 % at 200 W_e up to a maximum of 60 % at 1500 W_e, it then decreases to approximately 56 % at a 2000 W_e capacity. The thermal output from BlueGEN increases fairly linearly from 320 W_{th} at 200 W_e up to 540 W_{th} at 1500 kW_e. The thermal output increase is then much steeper, up to a maximum of 1000 W_{th} at 2000 W_e. At the optimised 1500 W_e output a CHP efficiency of 81.6 % is achieved.

The field trial data shown in Fig. 7.3 demonstrates a net electrical efficiency of 60 % is achievable and maintainable at a 1.5 kW_e output in a real life building application and thus the data presented in Fig. 7.4a can be used with confidence.

Figure 7.4b shows the flow water temperature in the SOFC WHR loop as a function of electrical power output. The flow water temperature is calculated based on the thermal output data presented in Fig. 7.4a, a 2 L min^{-1} water volumetric flow and a 45 °C return water temperature. The flow water temperature ranges between 47 °C at 100 W_e electrical power output up to a maximum of and 52 °C at a 2000 W_e electrical power output. As highlighted in Chaps. 4 and 6, it is primarily the desiccant systems operation that needs to be optimised to facilitate effective tri-generation system integration. Using the WHR flow water temperature and SDCS empirical data shown in Fig. 6.16c, the COP_{th} of the SDCS can be

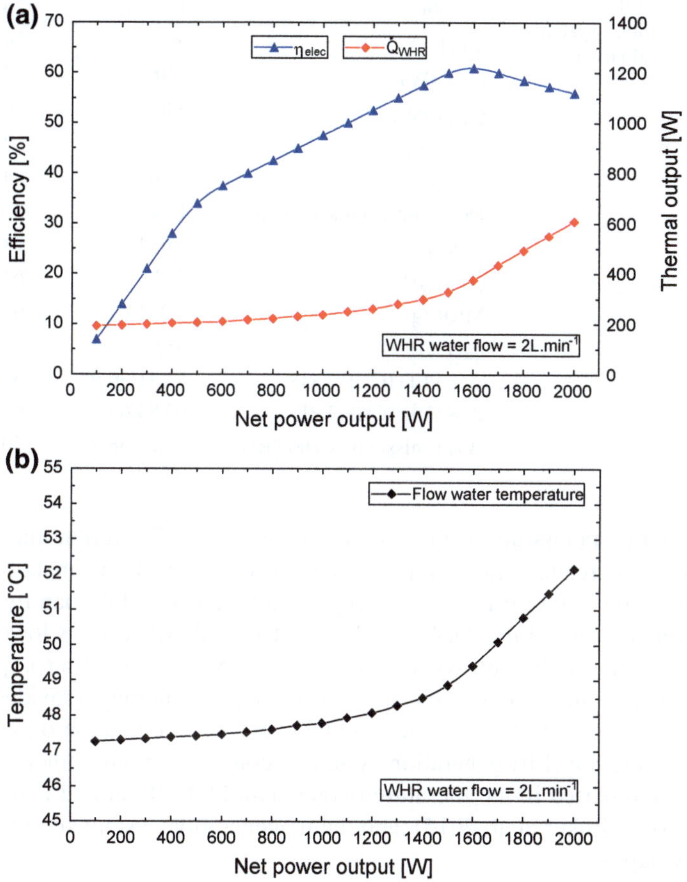

Fig. 7.4 **a** SOFC electrical efficiency and thermal output, and **b** WHR flow temperature as a function of electrical output (Foger 2013)

identified, and the cooling output calculated. Table 7.1 presents the results from the theoretical integration of the BlueGEN SOFC and SDCS into a complete tri-generation system at a net 1.5 and 2 kW_e output, operating with a 30 °C and 70 % RH inlet air condition. The parasitic energy consumption (110 W) of the SDCS has been included in the evaluation. The constants used for the PED, cost and emission analysis can be referred to in Table 4.6.

The theoretical integration study based on empirical data, demonstrates that high tri-generation efficiency in the range of 68–71 % is attainable when combining SOFC and liquid desiccant air conditioning technology; values in good agreement with the simulations presented in Chap. 4. The SOFC unit has a low heat to power ratio, particularly at the 1.5 kW_e condition, this is because it is an electrically optimised device (fuel utilisation of ~85 %). As a result, there is limited thermal output available for desiccant solution regeneration. However, the SDCS

Table 7.1 BlueGEN SDCS tri-generation system performance evaluation

Variable	1.5 kW_e	2 kW_e
η_{elec} (%)	60	56
$\dot{Q}_{CH_4}$ (W)	2500	3571
$\dot{Q}_{WHR}$ (W)	540	1000
$T_{WHR,flow}$ (°C)	48.87	52.16
η_{CHP} (%)	81.6	84
Desiccant volume (L min^{-1})	1.74	3.16
COP_{th}	0.614	0.649
$\dot{Q}_{cooling}$ (W)	332	649
MRR (g s^{-1})	0.2515	0.2941
η_{tri} (%)	68.9	71.1
Δ% PED (CHP/TRI)	51.41/46.98	50.21/46.79
Δ% Cost (CHP/TRI)	62.84/60.67	61.53/60.53
Δ% Emissions (CHP/TRI)	51.21/68.96	50.01/68.26

operating with a potassium formate solution at a 0.65–0.7 solution mass concentration has a low regeneration temperature requirement, and thus makes good use of the low-grade SOFC WHR output to generate a meaningful quantity of dehumidification/cooling. At the 2 kW_e condition, electrical efficiency is lower, but the thermal efficiency is higher. As a result, almost 650 W of cooling is produced. The inclusion of liquid desiccant air conditioning technology provides an efficiency increase of 9–15 % compared to SOFC electrical operation only. The performance of the novel tri-generation system is competitive with other systems of this capacity reported in the literature (Kong et al. 2005; Jradi and Riffat, 2014c; Wu et al. 2014; Buker et al. 2015), and the tri-generation system simulations presented in Chap. 4.

Table 7.1 shows that CHP and tri-generation efficiency is highest for the 2 kW_e case. However the PED, cost and emission savings are highest for the 1.5 kW_e case. Electricity has a higher associated cost and emissions compared to natural gas, therefore greater savings are made for the 1.5 kW_e case due to the higher electrical efficiency. In tri-generation cooling mode, relative cost and emission reductions compared to a conventional separated system for the 1.5 and 2 kW_e cases are around 60 and 70 % respectively, demonstrating the potential of the first of its kind SOFC liquid desiccant tri-generation system for building applications. A more comprehensive economic and environmental assessment is provided in Chap. 8. The operational issues encountered with the BlueGEN SOFC illustrate the real challenge of fuel cell deployment in the built environment. Reliability, durability and cost currently pose a great barrier to their wider use. Not until these issues are addressed will the operational advantages of fuel cells operating in the built environment be realised. Although experimental integration was not possible due to unforeseen circumstances, the stable nature prior to stack failure of the BlueGEN SOFC unit highlights the potential for the development of an efficient and effective tri-generation system.

Section 7.2 has presented a theoretical tri-generation system integration study based on empirical data. The work has established that SOFC and liquid desiccant are a viable technological pairing in the development of an efficient and effective tri-generation system. It has been demonstrated that high tri-generation system efficiency is attainable at low system capacities. The encouraging performance is primarily due to the high electrical efficiency of the SOFC and the reasonable COP_{th} of the liquid desiccant system. The proof-of-concept study has achieved two of the technical objectives of the thesis, namely a 1.5 kW_e system operating at an electrical efficiency of 45 % or higher. Meeting the overall system efficiency objective would require a higher desiccant system COP_{th}.

Next, Sect. 7.3 presents the experimental tri-generation system, system configuration, experimental set-up, instrumentation and experimental method.

7.3 Experimental Tri-generation System

Figure 7.5 shows a schematic diagram of the experimental SOFC liquid desiccant tri-generation system. The system is divided into three main components:

(1) SOFC power generation
(2) SOFC WHR circuit
(3) Liquid desiccant air conditioning.

A 250 W_e micro-tubular SOFC has been acquired by The University of Nottingham for tri-generation integration, and is designed to operate with propane (C_3H_8) as a

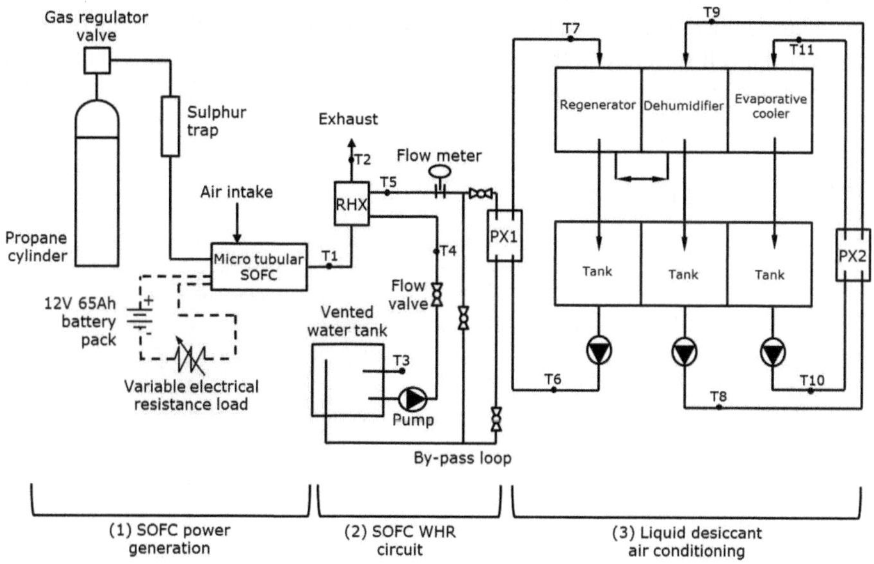

Fig. 7.5 Experimental tri-generation system schematic

fuel. The SOFC WHR system was developed for this project by The University of Nottingham. As a result actual thermal output values for the unit were not known before testing; however the manufacturer estimated that 500–1000 W was available for heat recovery. The SDCS, documented in Chap. 6, is employed as the thermally activated cooling technology in the experimental tri-generation system.

The proposed tri-generation system is truly multi-functional, and can provide electrical power and heating through the SOFC CHP unit, in addition to dehumidification and cooling/thermal comfort control through the liquid desiccant air conditioning unit. The novel tri-generation system is well suited to hot and humid climates, and applications that require onsite electrical power and simultaneous heating or dehumidification/cooling. Next, Sect. 7.3.1 describes the micro-tubular SOFC CHP component.

7.3.1 Micro-tubular SOFC CHP Component

The original, intended BlueGEN SOFC CHP system is of a planar cell type structure. The newly acquired SOFC CHP unit is of a micro-tubular cell type structure. An introduction to micro-tubular SOFC technology is provided below.

Unlike other fuel cells, SOFCs can have multiple cell geometries. There are three SOFC geometries: planar, co-planar and micro-tubular. The planar type cell is currently the most commonly used geometry, seen in units such as the BlueGEN, and was described in detail in Chap. 4.

The micro-tubular type SOFC was invented in the early 1990s by Kevin Kendall (Howe et al. 2011). Most commonly, the materials used (Ni-YSZ/LSM/YSZ) and operating temperatures (600–1000 °C) are similar to those of the planar type cell however the cell structure is that of a hollow extruded tube of 1–5 mm in diameter, sealed at one end. The structure of this design is shown in Fig. 7.6a and can either

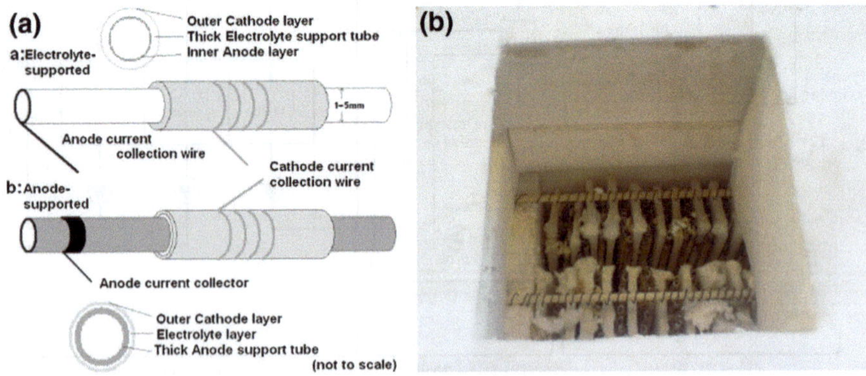

Fig. 7.6 **a** Basic SOFC micro-tubular designs (Howe et al. 2011), and **b** a 100 tube SOFC stack for CHP demonstration

be electrolyte or anode supported. Depending on the support design, air or fuel is passed through the inside of the tube and the other gas is passed along the outside of the tube. The support tube is longer than the active cell length. The first tube segment provides a gas inlet tube and the outlet section can be used as a combustor tube, where the fuel (hydrogen or short chain hydrocarbons) and oxidant (oxygen or air) combine. Nickel and silver current collection wires are wound around the outside of the tubes, attached to the anode and cathode tubes respectively.

Because each tube only produces 0.5–5 W of electrical power, the single tubes are bundled together in various configurations to form stacks to meet application demands. Figure 7.6b shows a 100 tube stack designed by IVF-Swerea for a CHP demonstration project. A scale-up issue does exist for large power demand applications. This is due to the challenges of a large number of electrical connections and gas inlets, which require further equipment compared to the planar type design. In a CHP system, an integrated afterburner/recuperator is employed to combust unconsumed fuel leaving the stack and pre-heat the inlet cathode airstream. The hot gases exiting the afterburner/recuperator are then passed through a second recuperator heat exchanger, transferring thermal energy to a working fluid such as water, to be used in a useful process.

Micro-tubular SOFCs show desirable operational characteristics such as high volumetric power density, good endurance against thermal cycling, and flexible sealing between fuel and oxidant streams (Panthi and Tsutsumi 2014). Furthermore, rapid start-up is possible with the micro-tubular design, one minute has been demonstrated, contrasting to the 1–24 h start-up times more conventionally observed with planar type stacks. As a result, micro-tubular SOFCs are suitable for the requirement of frequent restarting if necessary. However, the performance, namely fuel utilisation and thus electrical efficiency of the micro-tubular design is low. This is because of high cell resistance.

Because of micro-tubular SOFCs favourable power densities and rapid start-up times, they are being developed for small portable chargers and unmanned aerial vehicle applications as well as the stationary market. The planar type stack design still holds the largest market share for built environment CHP applications; which is primarily due to higher electrical efficiency and more favourable heat to power ratios. Currently, lower operating temperatures and higher power densities are the two main research targets for both planar and micro-tubular SOFC technology.

The micro-tubular SOFC CHP system has been acquired from Adelan Ltd. and installed in the laboratories at The University of Nottingham. The micro-tubular SOFC CHP test rig consists of the 250 W_e micro-tubular SOFC unit, a propane cylinder, gas regulator valve, sulphur trap, 12 V DC battery pack, a 12 V 250 W variable electrical load and a WHR circuit. A labelled photograph of the SOFC CHP experimental set-up is shown in Fig. 7.7.

The micro-tubular SOFC unit is made up of 48 anode supported co-extruded Ni-YSZ/LSM/YSZ tubes grouped together in bundles of four. These bundles are connected in series to form the stack. Each tube is 10 cm long, with a 20 cm^2 surface area, and can produce 5–10 W of DC electrical power. The start-up time is 10–20 min, and can endure 3000 h of operation with 100 on/off cycles.

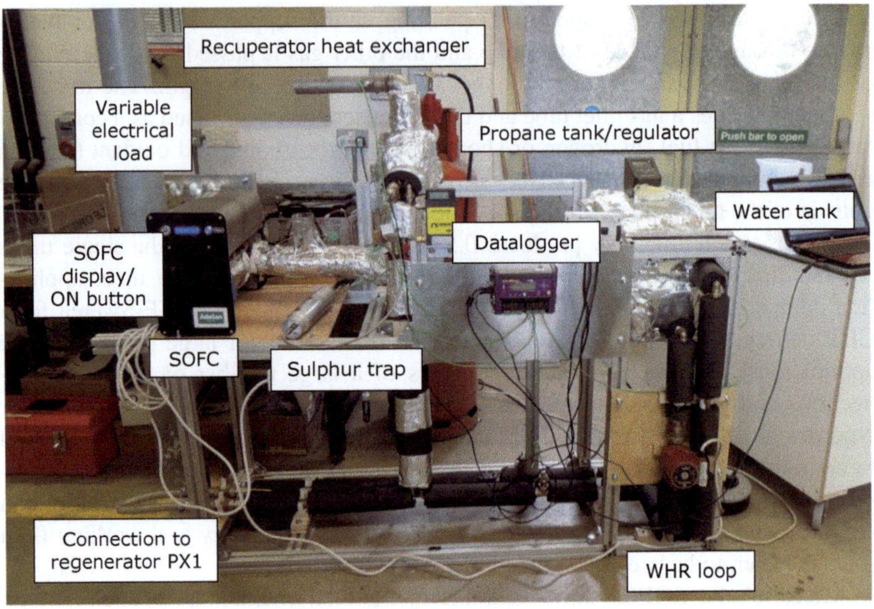

Fig. 7.7 Experimental test set-up for the micro-tubular SOFC CHP unit, with propane bottle, sulphur trap, electrical discharger and WHR circuit

A 50 L 2.5 bar propane tank is connected to the rear of the SOFC unit via a 1 bar gas regulator valve and sulphur trap. An advantage of SOFCs is there resilience to fuel poisoning, however they are still vulnerable to sulphur poisoning, and thus a sulphur trap is necessary. The propane regulator is set to 1 bar i.e. the fuel is supplied at atmospheric pressure. The micro-tubular SOFC unit uses approximately 100 g of propane per hour (1288 W fuel input). Ambient air is drawn in through the top of the unit via two 12 V DC axial flow fans. One airstream is delivered to the front of the unit, to be mixed with the propane fuel, at an air to fuel ratio of 3:1. The other airstream is passed through the unit's recuperator/afterburner to be pre-heated and then delivered to the cathode side of the cell to facilitate the electrochemical reactions. The propane air fuel mixture is introduced to the inside of the tubular cells where it is first heated to ~600 °C and catalytically converted to hydrogen and carbon monoxide. The fuel is then delivered to the anode, where the electrochemical reactions take place, generating electricity, heat and water vapour. The stack has a fuel utilisation factor of 40–50 %, typical of a micro-tubular type stack. The parasitic power consumption of the fans is 35 W. The micro-tubular SOFC systems 12 V DC electricity output is connected to two parallel connected 12 V 65 Ah battery packs, which in turn are connected to an array of five 50 W 12 V lights.

Following the electrochemical reactions, the hot gases (hydrogen, propane, carbon monoxide, carbon dioxide and water vapour) leaving the micro-tubular SOFC stack are combusted in the afterburner. An integrated recuperator serves to pre-heat the cathode air inlet. The system is thermally managed by adjusting the cathode air flow to the stack. The hot gases leaving the afterburner are at approximately 350 °C. These hot gases are directed into a stainless steel flue, where they are then passed through a gas to liquid recuperator heat exchanger. Here the thermal energy is transferred to a water working fluid in the WHR circuit. The WHR circuit is made of 22 mm copper pipe lagged with 19 mm Climaflex Pipe Insulation. A Wilo-Smart A-rated 230 V AC pump is used to circulate the hot water in the heating circuit. A 22 mm gate valve is fitted to the WHR circuit to control the water volumetric flow. The WHR loop is connected to a vented 30 L insulated water tank. During CHP tests, a by-pass loop is used to circulate water from the recupertaor heat exchanger to the tank. During tri-generation testing the flow valves are set so that the hot water is directed to PX1 to heat the desiccant solution in the SDCS. Following waste heat recovery, the SOFC exhaust gases are rejected to the environment. Based on a 250 W_e output and the LHV of propane, the net electrical efficiency of the micro-tubular SOFC is approximately 19.4 %.

Section 7.3.1 has described the micro-tubular SOFC CHP component. Next, Sect. 7.3.2 details the liquid desiccant air conditioning component.

7.3.2 Liquid Desiccant Air Conditioning Component

The thermally activated cooling technology used in the tri-generation system is the SDCS documented in Chap. 6. The SDSC employs a potassium formate solution at a solution mass concentration of 0.65–0.7. A comprehensive SDCS rig schematic and photograph is provided in Figs. 6.1 and 6.3 respectively, along with rig description, experimental set-up, instrumentation, and experimental testing method. Therefore it will not be repeated again, only with respect to the complete tri-generation system testing. The experimental operation of the SDCS in a tri-generation system context is provided in Sect. 7.3.4, and the SDCS operational values are provided in Sect. 7.4.2.

Next, Sect. 7.3.3 describes the instrumentation used in the experimental tri-generation system.

7.3.3 Instrumentation

This section describes the instrumentation used in the experimental tri-generation system. The tri-generation system is made up of the micro-tubular SOFC CHP and liquid desiccant components, therefore their instrumentation are described accordingly.

In the micro-tubular SOFC CHP system, the electrical power output is determined by measuring the voltage and current output. The operating voltage is displayed on the SOFC display panel and the result recorded. The current is measured using a GMC-I CP41 current clamp meter. This is placed over the positive wire connecting the micro-tubular SOFC to the battery pack, and the result recorded. K-Type thermocouples have been placed in the SOFC exhaust gas stream before (T1) and after (T2) the recuperator heat exchanger. The water volumetric flow in the WHR circuit is measured using an RS 1–15 L min^{-1} piston flow meter. K-Type thermocouples are placed on the WHR circuit at the recuperator heat exchanger inlet (T4) and outlet (T5), and in the water tank (T3). All micro-tubular SOFC thermocouples are connected to a DT80 DataTaker datalogger. Data readings are recorded every 10 s. The SDCS employs Type-K thermocouples on all desiccant solution and water flows (T6–T11). All inlet and outlet air flows have been instrumented with Vaisalia HMP110 humidity and temperature probes. A DataTaker DT500 datalogger is used to record data from the thermocouples and humidity and temperature probes every 10 s. All outlet air velocities are measured using an RS AM4204 hot wire anemometer. The density of the desiccant solution is measured using a differential pressure density meter with temperature compensation. Further information regarding the SDCS instrumentation can be referred to in Sect. 6.2.1. Details of the tri-generation systems instrumentation and associated accuracy are listed in Table 5.1. Section 7.3.3 has described the instrumentation used in the experimental tri-generation system (SOFC and SDCS). Next, Sect. 7.3.4 describes the experimental method used in the experimental testing and evaluation of the micro-tubular SOFC CHP and novel tri-generation system.

7.3.4 Experimental Method

To start the micro-tubular SOFC CHP system, the gas valve on the propane cylinder is opened. The micro-tubular SOFC electrical output cables are then connected to the battery pack. The micro-tubular SOFC ON button is then pressed. The digital display on the micro-tubular SOFC unit will show the operating voltage of the battery pack and the number of hours of operation the unit has completed to date. The voltage of the battery pack needs to be at 11.8 V for the micro-tubular SOFC unit to begin start-up. If the voltage of the battery pack is greater than this, the electrical load needs to be connected to discharge the battery. Once 11.8 V is achieved the micro-tubular SOFC will go into heat-up mode. Heating of the micro-tubular SOFC system is achieved through the combustion of propane in the afterburner. This takes approximately 20 min. During this time a current flow of 2.95 Å from the battery pack to the micro-tubular SOFC is observed: this is due to the micro-tubular SOFC parasitic energy consumption (35 W). Once the system is up to temperature (~600 °C), it goes into power production. Now a current flow of approximately 20 Å (when operating at a 250 W_e output) from the micro-tubular SOFC to the battery pack is observed. It is important that the electrical load (lamps) on the battery pack is maintained during micro-tubular SOFC operation to avoid the battery voltage exceeding 13 V and the micro-tubular SOFC shutting down. Voltage and current readings are taken every two minutes, and the results recorded. The sulphur trap has a lifetime of 250 h. It is essential this is not exceeded as sulphur poisoning will damage the stack. The micro-tubular SOFC unit's operating hour counter is on a 250 h loop so that the replacement milestones are clear. Throughout micro-tubular SOFC CHP and tri-generation system tests, a constant fuel input of 100 g h^{-1} is assumed.

During the micro-tubular SOFC heat up period, the pump in the WHR circuit is switched on and the water volumetric flow is set to the required test conditions. During micro-tubular SOFC CHP tests, the water in the WHR is simply circulated from the tank and through the recupertaor plate heat exchanger using the by-pass loop. The WHR inlet and outlet water temperatures are recorded and used to calculate the WHR thermal output. The tank volume is sufficient to act as a thermal load to the micro-tubular SOFC. During tri-generation system testing the micro-tubular SOFC CHP system is operated in the same manner as described above. However, the water in the WHR loop is diverted through the regenerator plate heat exchanger (PX1) to heat the desiccant solution. The SDCS experimental method has been provided previously in detail in Sect. 6.2.2. The experimental metrics used to evaluate the performance of the micro-tubular SOFC CHP and tri-generation system are provided below.

The DC electrical power output of the micro-tubular SOFC is determined using Eq. 7.1.

$$\dot{W}_{elec,DC} = I \times V \tag{7.1}$$

The thermal output of the micro-tubular SOFC is determined using Eq. 7.2.

$$\dot{Q}_{WHR} = \dot{m}_{w,WHR} c_{p,w,WHR} \left(T_{WHR,out} - T_{WHR,in} \right) \tag{7.2}$$

The heat capacity is evaluated using a validated EES function. The electrical efficiency of the micro-tubular SOFC is calculated using Eq. 7.3, and is based on the LHV of the propane fuel (4.635×10^7 J kg^{-1}). The mass flow rate of the fuel input is assumed constant at 0.00002778 kg s^{-1} (100 g h^{-1}).

$$\eta_{elec} = \frac{\dot{W}_{elec,DC}}{\dot{m}_{fuel} \cdot LHV_{C_3H_8}} \tag{7.3}$$

The co-generation efficiency of the micro-tubular SOFC CHP system is determined using Eq. 7.4.

$$\eta_{CHP} = \frac{\dot{W}_{elec,DC} + \dot{Q}_{WHR}}{\dot{m}_{fuel} \cdot LHV_{C_3H_8}} \tag{7.4}$$

The complete tri-generation system efficiency is shown in Eq. 7.5.

$$\eta_{tri} = \frac{\dot{W}_{elec,net} + \dot{Q}_{WHR} + \dot{Q}_{cooling}}{\dot{m}_{fuel} \times L\dot{H}V_{C_3H_8}} \tag{7.5}$$

During the tri-generation system evaluation, $\dot{W}_{elec,net}$ is the DC electrical output of the micro-tubular SOFC minus the parasitic energy consumption of the SDCS ($\dot{W}_{aux,des}$).

Section 7.3.4 has provided the experimental method and evaluation metrics used to assess the micro-tubular SOFC CHP and tri-generation system. Section 7.4 presents the experimental results and analysis.

7.4 Experimental Tri-generation System Results and Analysis

The performance of the SDCS is documented in detail in Chap. 6, in which the regenerator thermal input values (water volumetric flow and water flow temperature) were selected based on a range that matched the 1.5 kW_e (BlueGEN) SOFC CHP system. However, due to the BlueGEN SOFC not being available for tri-generation system integration, a 250 W_e (Adelan) micro-tubular SOFC unit had to be acquired. Actual thermal output values for this unit were not known before testing because the micro-tubular SOFC WHR system was developed specifically for this project by The University of Nottingham. Prior to this the micro-tubular SOFC had only been used for electrical production. Section 7.4.1 presents micro-tubular SOFC CHP system component analysis, which aims to quantify the thermal performance of the SOFC before tri-generation system integration and analysis in Sect. 7.4.2.

7.4.1 Micro-tubular SOFC CHP System Component Analysis

In this section the micro-tubular SOFC CHP system component test results are presented and analysed. The results are provided to show (a) the performance of the micro-tubular SOFC unit (electrical and thermal) over time and (b) to characterise the thermal performance to facilitate effective tri-generation system integration in Sect. 7.4.2. The micro-tubular SOFC CHP unit operates at a constant output. The input fuel flow rate is fixed and thus so power and thermal output is approximately constant. The water volumetric flow in the WHR circuit is set to 2 L min^{-1} to replicate the SDCS testing conditions presented in Chap. 6.

Before being supplied to The University of Nottingham, long term stability testing of the micro-tubular SOFC was carried out at The University of Birmingham.

This testing data is presented in Appendix 4. The micro-tubular SOFC was run for 130 h, with 19 thermal cycles. A steady-state electrical output of 250 W has been demonstrated. However, after 75 h of operation severe sulphur poisoning of the micro-tubular SOFC stack occurred. The sulphur trap had been previously used and went over the 250 h operating limit. In the hours following sulphur trap replacement very little power output could be gained from the micro-tubular SOFC. However, after approximately 95 h of operation, the micro-tubular SOFC had recovered to a final maximum power output of 140 W. The micro-tubular SOFC has now recovered a little more to a 150.4 W electrical output, which is approximately 60 % of the manufacture's quoted 250 W capacity. Sulphur poisoning and the regeneration of Ni-based anodes in SOFCs has been investigated by Zha et al. (2007). The degradation in cell performance is attributed to rapid adsorption of sulphur onto the Ni surface to form nickel sulphide, which blocks the active sites for hydrogen adsorption and oxidation. Following removal of H_2S (hydrogen sulphide) from the fuel stream, the anode performance can, depending on operating conditions and duration of H_2S exposure, recover fully or partially. The rate of the recovery process increases with operating temperature and cell current density.

Figure 7.8 shows the electrical and thermal performance of the micro-tubular SOFC CHP unit over a 350 min test period. The first 28 min shows the micro-tubular SOFC heat-up. During this period the micro-tubular SOFC has a 35 W parasitic load on the batteries, however there is a thermal output. Following heat-up, the micro-tubular SOFC goes into power production. During this period there is an electrical and thermal output. At 322 min, the micro-tubular SOFC is turned off, and it takes 22 min to cool down.

Figure 7.8a shows that during the power production period (minute 28–322) the micro-tubular SOFC produces an average of 150.4 W of DC electrical power (12.2 V at a current flow of 12.33 Å). This equates to an electrical efficiency of 11.68 %, in comparison to 19.4 % at the original 250 W electrical output.

Figure 7.8b shows the inlet and outlet water temperatures in the WHR circuit. At a 2 L min^{-1} water volumetric flow, a maximum WHR outlet water temperature of up to 65 °C is possible, demonstrating the potential for desiccant solution regeneration in a tri-generation system context. Over the power production period, the average water temperature difference across the recuperator heat exchanger is 3.2 °C, this equates to an average thermal output of 446.9 W. The pinch temperature shown in Fig. 7.8b is the difference between the outlet flue gas temperature and outlet WHR water temperature. The pinch temperature reaches a peak of 20.01 °C at 70 min then gradually decreases to 11.53 °C at 322 min. The decline in pinch temperature is because the rate of increase in flue gas outlet temperature over time is greater than the rate of increase in the WHR outlet water temperature. This indicates a reduction in thermal energy extraction.

Based on the averaged values over the power production period, the micro-tubular SOFC achieved a CHP efficiency of 46.39 %. This compares with 54.1 % based on the micro-tubular SOFC unit before sulphur poisoning, which had an average power output of 250 W_e. The stable operational nature of the

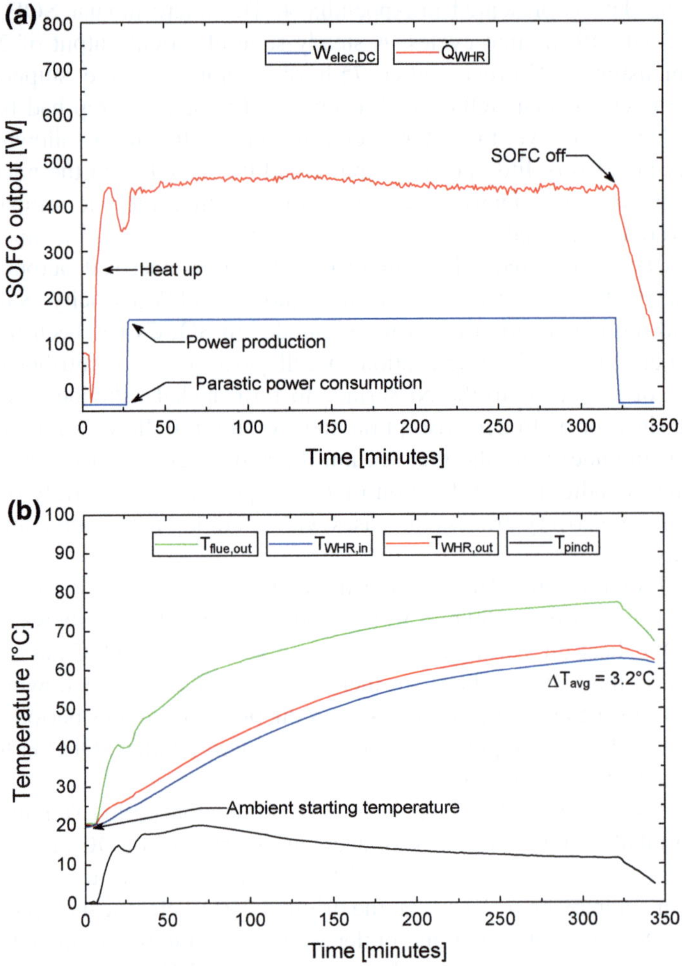

Fig. 7.8 **a** Micro-tubular SOFC CHP electrical and thermal output, and **b** WHR inlet/outlet, flue gas outlet and pinch temperatures

micro-tubular SOFC CHP unit demonstrates the potential for tri-generation system integration. Using Eq. 5.3, the maximum calculated relative uncertainties in the SOFC $\dot{Q}_{WHR}$ and η_{CHP} are ±9.1 and ±6.8 % respectively.

In comparison to other combustion based micro-CHP technologies of this electrical capacity, the electrical efficiency of the micro-tubular SOFC is reasonable. Compared to planar type SOFC systems, the micro-tubular SOFC has a low electrical efficiency. However, the significant advantage of micro-tubular SOFC technology has been confirmed. Quick start-up and shut-down times of 20 min have been demonstrated, meaning the unit can respond to rapid supply and demand requirements. The company (Adelan Ltd.) supplying the micro-tubular

SOFC have not previously attempted to provide WHR, and thus the WHR system was developed for this project by the University of Nottingham. The low thermal output from the micro-tubular SOFC highlights the need for future work on optimising and refining the WHR provision in order to maximise the thermal output and thus elevate the system efficiency. Future work should aim to improve the connection between the afterburner outlet and recupertaor heat exchanger flue inlet. As demonstrated in Chap. 6, a regenerator thermal input of less than 500 W will result in limited regeneration capacity. In order to maintain balanced liquid desiccant system operation, a restricted cooling output will have to be assigned to the dehumidifier.

Section 7.4.1 has presented component testing of a micro-tubular SOFC CHP unit. The micro-tubular SOFC was acquired at short notice to replace the building-installed 1.5 kW_e BlueGEN SOFC. Due to sulphur poisoning the micro-tubular SOFC unit has suffered a 40 % drop in electrical output from 250 to 150 W. Water flow temperatures in the WHR circuit of up to 65 °C at a 2 L min^{-1} water volumetric flow have been demonstrated. The micro-tubular SOFC has a low thermal output of approximately 450 W, thus it is not the ideal match for the developed SDCS and it is anticipated that regeneration capacity in the tri-generation system will be limited. However, the novel concept of integrating SOFC and liquid desiccant technology is still successfully demonstrated in Sect. 7.4.2

Next, Sect. 7.4.2 presents the results and analysis from tri-generation system integration and testing.

7.4.2 Tri-generation System Analysis

In this section the micro-tubular SOFC is integrated alongside the SDCS to form the complete tri-generation system. As previously highlighted, the micro-tubular SOFC operation is fixed, it has no provision to modulate or alter output. Therefore, it is the liquid desiccant systems operation that is controlled in order to investigate tri-generation system performance.

In the tri-generation system the connection between the micro-tubular SOFC and liquid desiccant components is through the regenerator. As a result of this arrangement, the tri-generation system analysis evaluates the performance of the regenerator at three different desiccant solution flows using the micro-tubular SOFC thermal input. This is to determine the instantaneous performance of the novel system. Following this, a daily tri-generation performance analysis is presented which serves to demonstrate the novel system operating in a nonsynchronous fashion in a building application.

For efficient and effective tri-generation system operation, a moisture balance between the dehumidifier and regenerator is required. As a result, the achievable regenerator moisture addition rate presented in this section is equated, using the data presented in Sect. 6.3.1, to a suitable dehumidifier moisture removal rate.

From the dehumidifier moisture removal rate, the achievable cooling output can be obtained, and the tri-generation system efficiency calculated. In Chap. 6, the SDCS dehumidifier was tested in the environmental chamber to simulate real life operating conditions. Thus the tri-generation system results presented are representative of the novel system in a real working environment.

Testing of the tri-generation system is carried out by operating the micro-tubular SOFC CHP system until an outlet water temperature of 50 °C is achieved with a 2 L min^{-1} water volumetric flow in the WHR circuit. This takes approximately 120 min from ambient. The by-pass loop is then closed and the hot water is directed to the regenerator desiccant solution plate heat exchanger (PX1). The regenerator desiccant solution and air flow is then turned on. The regenerator tests last for 90 min or until steady-state output data is achieved. The micro-tubular SOFC and regenerator are then turned off.

Three desiccant solution volumetric flows have been investigated: 1.2, 2.2, 3.2 L min^{-1} at a potassium formate desiccant solution mass concentration of 0.65–0.7. The regenerator volumetric air flow used is 256 $m^3 h^{-1}$. The aim of the investigation is to determine the conditions, at which the regenerator moisture addition rate is highest, and thus the instantaneous cooling output can be maximised. As previously highlighted, during tri-generation system testing the regenerator uses ambient laboratory air.

Figure 7.9 shows the performance of the regenerator operating on the thermal input from the micro-tubular SOFC. The plot shows the regenerator moisture addition rate and inlet solution temperature with respect to regenerator solution volumetric flow. As demonstrated in Figs. 6.12a and 6.13a, it is apparent that the regenerator moisture addition rate is related to both the desiccant solution volumetric flow and temperature. Operating with the micro-tubular SOFC thermal input, the highest regenerator moisture addition rate of 0.11 g s^{-1} is achieved at a 2.2 L min^{-1} desiccant solution flow. This desiccant solution volumetric flow

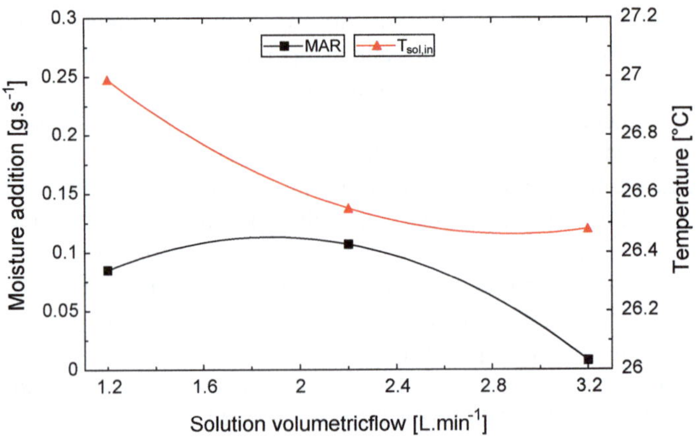

Fig. 7.9 Tri-generation system regenerator performance

Table 7.2 Operating values for experimental tri-generation system evaluation

Variable	Value	Variable	Value
SOFC fuel flow (g h^{-1})	100	Deh air temperature (°C)	30
WHR flow (L min^{-1})	2	Deh air relative humidity (%)	70
Reg $\omega_{a,in}$ (kg_{vapour}/kg_{dryair})	0.006412	Deh air vol. flow (m^3 h^{-1})	102
Reg air flow (m^3 h^{-1})	256	Deh des flow (L min^{-1})	3.2
Reg des flow (L min^{-1})	2.2	Des mass concentration (%)	65–70

achieves a balance between the volume and temperature of solution passing through the regenerator HMX. As a result, a 2.2 L min^{-1} solution volumetric flow has been selected for tri-generation system evaluation.

For successful tri-generation system evaluation a dehumidifier operating condition which balances the regenerator operation is required. Figure 6.7a shows a balanced (with the regenerator) dehumidifier moisture removal rate of 0.11 g s^{-1} is achieved when operating at a dehumidifier volumetric air flow of 102 m^3 h^{-1}. This equates to a cooling output of 278.6 W. Table 7.2 provides the operating values selected for tri-generation system evaluation.

Figure 7.10a, b show the respective steady-state performance of the micro-tubular SOFC and SDCS regenerator during tri-generation system testing. Figure 7.10a shows that the micro-tubular SOFC is operating in CHP mode up until 125 min. During this period 150.4 W of DC electrical power is produced with a thermal output of 418 W. At 128 min the WHR flow temperature reaches 51.76 °C, and the regenerator is turned on for a 90 min test. During the regenerator testing period, the WHR flow temperature drops dramatically, to eventually stabilise at approximately 32 °C. During the regenerator operating period the power output remains unchanged, however the WHR thermal output increases to 572.8 W. This increase in thermal output is due to the regenerator desiccant solution acting as a thermal load and thus lowering the return water temperature to the recupertaor heat exchanger. At 218 min, the micro-tubular SOFC and regenerator are turned off, and the system takes 22 min to cool down (Fig. 7.10).

Figure 7.10b shows the moisture addition rate in the regenerator and the inlet water temperature to regenerator PX1 over the 90 min test period. For the first 15 min the results are unstable. Following this, the regenerator moisture addition rate reduces in proportion with the inlet water temperature. As the desiccant solution is circulated through regenerator PX1 it is heated using the hot water in the SOFC WHR circuit. The low thermal output from the SOFC cannot maintain the flow temperature of 50 °C in the WHR circuit and as a result the water flow temperature decreases over time. Over the regenerator test period, the inlet water temperature decreases from an initial value of 50 °C to approximately 32 °C after 50 min when it becomes steady. The moisture addition rate of 0.11 g s^{-1}, shown in Table 7.3, is an average value taken between 60 and 90 min as this is a period of steady-state operation. The average solution temperature in this period is 27.2 °C.

Table 7.3 presents the instantaneous performance of the novel micro-tubular SOFC liquid desiccant tri-generation system. The performance evaluation is

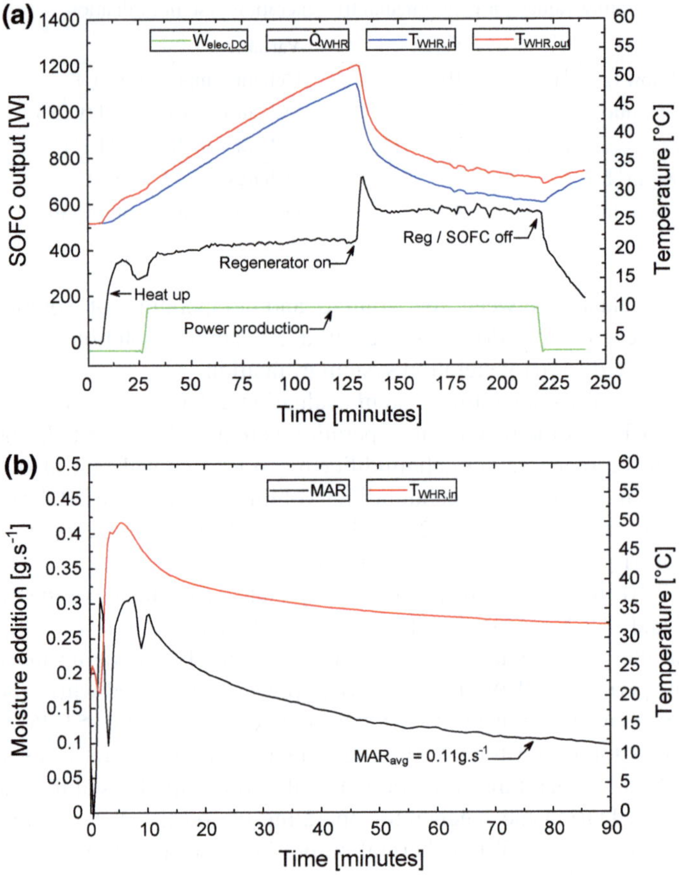

Fig. 7.10 Tri-generation system performance in the **a** micro-tubular SOFC, and **b** SDCS regenerator

Table 7.3 Instantaneous performance of the novel tri-generation system

Variable	Value	Variable	Value
$\dot{W}_{elec,DC}$ (W)	150.4	$\dot{Q}_{cooling}$ (W)	278.6
$\dot{Q}_{WHR}$ (W)	442.6	Deh_{MRR} (g s^{-1})	0.1114
$\dot{Q}_{C_3H_8}$ (W)	1288	Reg_{MAR} (g s^{-1})	0.11
η_{elec} (%)	11.68	COP_{th}	0.63
η_{CHP} (%)	46.04	η_{tri} (%)	24.77

provided at the 150.4 W electrical output. The 110 W parasitic energy consumption of the SDCS has been accounted for.

The marginal difference in the dehumidifier moisture removal and regenerator moisture addition rates is deemed insignificant enough for the purpose of tri-generation system evaluation. The novel system can generate 150.4 W of electrical power, 442.6 W of heat output or 278.6 W of cooling. Instantaneous tri-generation system efficiency is 24.77 %. At the original 250 W electrical output, the tri-generation system efficiency is 32.5 %. Without considering the parasitic energy consumption of the SDCS, the tri-generation system efficiency is 33.31 % at a 150.4 W electrical output and 41.04 % at the 250 W electrical output. When integrated with the micro-tubular SOFC, the SDCS demonstrates a COP_{th} of 0.62, an encouraging value for a waste heat driven cooling system of this capacity. Due to its low temperature regeneration requirement, potassium formate at a 0.65–0.7 mass concentration is an appropriate desiccant solution for a SOFC tri-generation system.

The tri-generation system efficiency is low. However, as highlighted in Sect. 7.4.1, the initial micro-tubular SOFC CHP system efficiency is below 50 %. Tri-generation system analysis shows that the low thermal output from the micro-tubular SOFC is insufficient to maintain a flow temperature of 45–50 °C and thus the regenerator moisture addition rate is low, resulting in a small instantaneous cooling output. Furthermore, almost all the micro-tubular SOFC electrical output is used for the parasitic energy consumption of the SDCS. However, the novel concept of integrating SOFC and liquid desiccant air conditioning technology into the first of its kind tri-generation system has been successfully demonstrated. The SOFC has been used to generate simultaneous electrical power, heating and dehumidification/cooling. The inclusion of liquid desiccant air conditioning technology provides an efficiency increase of up to 13 % compared to SOFC electrical operation only, demonstrating the merit of the novel tri-generation system in applications that require electricity, heating and dehumidification/cooling. Improvements to the micro-tubular SOFC WHR provision will improve tri-generation system performance.

As highlighted in Sect. 6.3.4, an operational advantage of a SOFC liquid desiccant tri-generation system is the potential for nonsynchronous operation. Re-concentration of the desiccant solution over extended time periods is an effective and efficient form of storing the constant thermal energy output from the SOFC with minimal losses. Furthermore, to meet a specific cooling load the dehumidifier can be operated at a higher cooling capacity, and the regenerator can be operated for an extended period to make up the moisture addition shortfall, creating a solution mass balance. Based on the assumption that the SOFC operates for 24 h a day, with a 6 h cooling period, Table 7.4 presents the daily tri-generation system performance, and serves to demonstrate the novel tri-generation system operating in a building application. The SDCS dehumidifier performance figures are taken from the steady-state operating values shown in Fig. 6.5.

The proposed daily tri-generation system operating concept demonstrates that the novel system can produce a peak cooling output of 527 W over a 6 h period. The daily tri-generation system efficiency is 37.9 %. At the original 250 W electrical output the daily tri-generation efficiency is 45.6 %. As the cooling period is increased the daily tri-generation efficiency decreases. This is because the SDCS has a COP_{th} of less than one. In this scenario the system would require the

Table 7.4 Daily tri-generation system performance

Variable	Value	Variable	Value
$\dot{W}_{elec,DC}$ (W)	150.4	Deh_{MRR} (g s^{-1})	0.21
$\dot{Q}_{WHR}$ (W)	442.6	Reg_{MAR} (g s^{-1})	0.11
$\dot{Q}_{cooling}$ (W)	527	Electrical energy (Wh)	3610
$\dot{Q}_{C_3H_8}$ (W)	1288	Heating energy (Wh)	5607
Electrical time (h:min)	24	Cooling energy (Wh)	3162
Cooling time (h:min)	6:00	Fuel input (Wh)	30,912
Regenerator time (h:min)	11:27	$\eta_{tri,day}$ (%)	37.91
Heating time (h:min)	12:33		

provision of sufficient desiccant solution storage in order balance the dehumidifier and regenerator operation. Continuous micro-tubular SOFC operation is a reasonable assumption in a (domestic) building application as the small electrical output can be used for base load applications (lights, standby etc.).

The micro-tubular SOFC can operate on natural gas, in such a scenario the novel tri-generation system generates a cost and emission reduction of 56 and 42 % respectively compared to a base case scenario of grid electricity, gas fired boiler and electrical driven VCS. The constants used for the emission and economic analysis can be referred to in Table 4.6. The encouraging economic and environmental performance demonstrates the potential of the novel tri-generation system in applications that require simultaneous electrical power, heating and dehumidification/cooling.

Section 7.4.2 has presented tri-generation system integration and analysis. The novel concept has been proven, experimentally, in the first of its kind system; however the reported performance is low. This is primarily due to the low thermal output from the micro-tubular SOFC. Possible solutions to improve performance have been discussed. Next, Sect. 7.5 presents the chapter conclusions with particular respect to the achievement of the thesis aim.

7.5 Chapter Conclusions and Achievement of the Thesis Aim

This section concludes the chapter, with particular respect to the achievement of the thesis aim: the design, development and testing development of an efficient and effective SOFC liquid desiccant tri-generation system.

This chapter has successfully presented the evaluation, based on experimental data, of a first of its kind SOFC liquid desiccant tri-generation system. Two SOFC liquid desiccant tri-generation systems have been presented. First, Sect. 7.2 presented a theoretical integration analysis, based on collected empirical SOFC CHP and liquid desiccant data. Second, an experimental tri-generation system has been introduced. Section 7.3 provides a summary of this novel experimental tri-generation system and Sect. 7.4 presents the results and analysis from the tri-generation system evaluation.

Technical and commercial issues have meant the 1.5 kW_e building-installed (BlueGEN) SOFC CHP system was not available for tri-generation system integration. However, using collected empirical SOFC and SDCS data, a theoretical tri-generation system integration analysis has been completed. The tri-generation system performance has been evaluated at a 1.5 and 2.0 kW_e capacity. The highest tri-generation efficiency of 71.1 % is achieved at a 2.0 kW_e capacity; however the electrical efficiency is lower than the 1.5 kW_e case. As a result, the 1.5 kW_e case produces the greatest cost and emission savings. The encouraging performance is primarily due to the high electrical efficiency of the SOFC and the reasonable COP_{th} of the liquid desiccant system. The inclusion of liquid desiccant air conditioning technology provides an efficiency increase of 9–15 % compared to SOFC electrical operation only. The performance of the novel tri-generation system is competitive with other systems of this capacity reported in the literature, and in good agreement with simulations presented in Chap. 4. The technical and commercial issues encountered with the 1.5 kW_e building-installed BlueGEN SOFC highlight the real challenge of fuel cell deployment in the built environment. Reliability, durability and cost currently pose a great barrier to their wider use, and demonstrate the need to focus future work on addressing these issues. However, the SDCS shows significant potential for integration with other CHP prime mover technologies such as ICE or SE. Due to the greater technological maturity of ICE and SE, the reliability of the complete tri-generation system can be expected to be improved.

Following the failure of the 1.5 kW_e SOFC, a 250 W_e micro-tubular SOFC had to be acquired. The novel tri-generation system concept has been proven experimentally using the micro-tubular SOFC and SDCS. The experimental results demonstrate regeneration of the potassium formate solution at a 0.65–0.7 solution mass concentration using the thermal output from the SOFC in the first of its kind tri-generation system. The novel system can generate 150.4 W of electrical power, 442.6 W of heat output or 278.6 W of cooling. Instantaneous tri-generation system efficiency is low at approximately 25 %. This is primarily due to the low capacity and poor performance of the micro-tubular SOFC. Insufficient WHR means only 450 W of thermal energy is available for regeneration purposes, and thus the cooling output is low. However, it has been suggested that if these issues are addressed, the novel system can provide higher overall efficiency.

The thesis has established that a clear operational advantage of the novel SOFC liquid desiccant tri-generation system is the potential for nonsynchronous operation. The constant SOFC thermal output can be used to re-concentrate the desiccant solution as a form of thermal energy storage. Unlike thermal storage techniques based on sensible energy, a significant advantage of (chemical) thermal energy storage in the form of strong desiccant solution is that there are minimal losses over time. Using this nonsynchronous operating concept, the experimental system can generate an increased peak cooling output of up to 527 W and a daily tri-generation efficiency of 37.9 %. This is an encouraging value for a tri-generation system of this capacity, and serves to demonstrate the novel tri-generation system operating in a building application. Compared to a base case scenario, the novel tri-generation system generates a cost and emission reduction of 56 and 42 % respectively, demonstrating the potential of the novel tri-generation system in applications that require simultaneous electrical power, heating and dehumidification/cooling.

The difference in performance seen between the two tri-generation systems presented in this chapter demonstrates the significance of (a) the performance of the SOFC component and (b) the requirement of optimal paring of components in the development of an efficient and effective tri-generation system. The micro-tubular SOFC was acquired at short notice to replace the 1.5 kW_e BlueGEN SOFC. As seen in the low thermal output, it is not the ideal match for the developed SDCS. However, the novel tri-generation concept has been successfully demonstrated. Both tri-generation system analyses presented have considered balanced liquid desiccant system operation, demonstrating the strength and rigour of the work presented.

Based upon the experimental work presented in this chapter, three conclusions are provided with respect to the design, development and testing of an efficient and effective tri-generation system based on SOFC and liquid desiccant air conditioning technology for building applications.

(1) SOFC and liquid desiccant is an effective technological paring. The inclusion of liquid desiccant can bring significant improvement to system performance, particularly in applications requiring simultaneous electrical power, heating and dehumidification/cooling.
(2) Overall tri-generation system performance is more influenced by the SOFC component than the liquid desiccant. Appropriate matching of component capacity is necessary.
(3) The novel tri-generation system concept has been demonstrated experimentally. Future work needs to focus on improving the current unreliability and sensitivity of fuel cell technology.

The aim of the thesis is to design, develop and test an efficient and effective tri-generation system based on SOFC and liquid desiccant air conditioning technology. This chapter has demonstrated a clear contribution to new knowledge with the development and evaluation of two SOFC liquid desiccant tri-generation systems and as a result it is proposed that the thesis aim has been completed.

Next, Chap. 8 provides an economic and environmental assessment of the novel SOFC liquid desiccant tri-generation system.

References

Buker, M.S., B. Mempouo, and S.B. Riffat. 2015. Experimental investigation of a building integrated photovoltaic/thermal roof collector combined with a liquid desiccant enhanced indirect evaporative cooling system. *Energy Conversion and Management* 101: 239–254.

Elmer, T., M. Worall, S. Wu, and S. B. Riffat. 2015a. Fuel cell technology for domestic built environment applications: State of-the-art review. *Renewable and Sustainable Energy Reviews* 42: 913–931.

Elmer, T., M. Worall, S. Wu, and S. Riffat. 2016a. Assessment of a novel solid oxide fuel cell tri-generation system for building applications. *Energy Conversion and Management* **124**: 29–41.

Foger, K. 2013. The technology of ceramic fuel cells. *Electrical Developments* 48–51.

Howe, K.S., G.J. Thompson, and K. Kendall. 2011. Micro-tubular solid oxide fuel cells and stacks. *Journal of Power Sources* 196(4): 1677–1686.

Jradi, M., and S. Riffat. 2014c. Experimental investigation of a biomass-fuelled micro-scale tri-generation system with an organic Rankine cycle and liquid desiccant cooling unit. *Energy* 71: 80–93.

Kong, X.Q., R.Z. Wang, J.Y. Wu, X.H. Huang, Y. Huangfu, D.W. Wu, and Y.X. Xu. 2005. Experimental investigation of a micro-combined cooling, heating and power system driven by a gas engine. *International Journal of Refrigeration* 28(7): 977–987.

Panthi, D., and A. Tsutsumi. 2014. Micro-tubular solid oxide fuel cell based on a porous yttria-stabilized zirconia support. *Scientific Reports* 4.

Sommer, K., and E. Messenholler. 2013. Practical experience with a fuel cell unit for Combined Heat and Power (CHP) generation on the building level. *REHVA Journal* 12.

Wu, J.Y., J.L. Wang, S. Li, and R.Z. Wang. 2014. Experimental and simulative investigation of a micro-CCHP (micro combined cooling, heating and power) system with thermal management controller. *Energy* 68: 444–453.

Zha, S., Z. Cheng, and M. Liu. 2007. Sulfur poisoning and regeneration of Ni-based anodes in solid oxide fuel cells. *Journal of the Electrochemical Society* 154(2): B201–B206.

Chapter 8
Economic and Environmental Assessment

8.1 Introduction

The aim of this chapter is to conduct an assessment as to whether the SOFC liquid desiccant tri-generation system is economically and environmentally viable. It builds upon the tri-generation system integration evaluation presented in Chap. 7 with an economic assessment in Sect. 8.2 and an environmental assessment in Sect. 8.3. Section 8.4 contextualises the presented results and concludes the chapter. A published journal article, by the author, on the economic and environmental assessment of the novel tri-generation system can be referred to here; Elmer et al. (2016a).

The economic and environmental assessments uses the BlueGEN SOFC tri-generation system performance data presented in Sect. 7.2, operating at a 1.5 and 2.0 kW_e capacity. These data, as opposed to the simulation data in Chap. 4 and laboratory testing data in Sect. 7.4, are used because the system demonstrates good performance and is based upon commercially available components which have long-term stable performance, accurate costing figures and qualify for government support.

The aim of the economic and environmental assessment is to establish:

1. Whether the proposed tri-generation system is economically and environmentally viable under current conditions compared to an equivalent base case system.
2. The conditions and geographical locations in which the novel tri-generation system is economically and environmentally viable compared to an equivalent base case system.
3. The future feasibility of the novel tri-generation system with respect to projected changes in global energy resources, conversion techniques and cost.

T. Elmer, *A Novel SOFC Tri-generation System for Building Applications*,
Springer Theses, DOI 10.1007/978-3-319-46966-9_8

8.2 Economic Assessment

In this section, an economic assessment of the novel SOFC liquid desiccant tri-generation system operating within a UK economic climate is presented. The economic assessment compares the 1.5 and 2.0 kW_e capacity tri-generation systems to an equivalent base case system comprising grid electricity, natural gas fired boiler and electrically driven VCS over a 15 year time period. The economic evaluation metrics used are: net present cost (NPC), equivalent uniform annual cost (EUAC) and simple pay-back period (SPBP). The unit cost of electricity, unit cost of natural gas and the capital cost of the SOFC are varied, in a reasonable range, to carry out a sensitivity analysis of the NPC and SPBP. Using electrical unit cost data published by the International Energy Agency (IEA 2012), the economic performance of the tri-generation system in the context of different countries is presented.

Next, Sect. 8.2.1 describes the metrics used in the economic assessment.

8.2.1 Economic Assessment Metrics

NPC, EUAC and SPBP are used to assess the economic performance of the novel SOFC liquid desiccant tri-generation system compared to a base case system.

8.2.1.1 Net Present Cost (NPC)

Net present value (NPV) is an economic tool used to equate the total cost of a project over a specified time period to the total cost today, taking into account the time value of money. The present value (PV) of each annual cash flow is discounted back to its PV using a suitable interest or discount rate. The NPV is determined by summing the PV for each year, staring at year 0 i.e. the investment, to the final year (N). NPV is a good indicator of how much value an investment or project brings to an investor, and is widely used in economic engineering to assess feasibility. However, there are many kinds of systems or projects, such as the SOFC tri-generation system, where there are no sales or incomes. In this case it is common to use net present cost (NPC). Equation 8.1 is used to calculate NPC (Pilatowsky et al. 2011).

$$\mathrm{NPC} = \sum_{t=0}^{N} \frac{AA_{TC}}{(1+i_r)^n} + I_{cc} \tag{8.1}$$

AA_{TC} is the adjusted annual total costs (£), i_r is the interest rate, n is the year number and I_{cc} is the initial capital cost (£). Selection of a suitable interest/discount rate is based upon risk, opportunity cost or an alternative investment.

In engineering based analysis 7 % is a widely used value. If inflation is being considered, the adjusted annual total cost (AA_{TC}) is calculated using Eq. 8.2.

$$AA_{TC} = A_{TC}\left(1 + i_f\right)^n \tag{8.2}$$

A_{TC} is the non-adjusted annual total costs (£), i_f is the inflation rate and n is the year number. The scrap value (SV) of the system at the end of the project's life should be considered, and subtracted from the final expenditure. In NPC analysis the annual total expenditure or costs (AA_{TC}) are given as positive figures (unlike NPV), and thus the NPC at the end of a system lifetime will be positive. When two or more systems are being evaluated over the same time period, the system with the lowest NPC should be selected.

8.2.1.2 Equivalent Uniform Annual Cost (EUAC)

The equivalent uniform annual cost (EUAC) is the annual cost of the project or system equivalent to the discounted total cost or NPC. EUAC is calculated by multiplying the NPC by the capital recovery factor (CRF), as shown in Eq. 8.3.

$$\text{EUAC} = \text{NPC}\left[\frac{i_r(1 + i_r)^n}{(1 + i_r)^n - 1}\right] \tag{8.3}$$

8.2.1.3 Simple Pay-Back Period (SPBP)

The simple pay-back period (SPBP) is used to determine the time required to recoup the funds expended in an investment, or to reach the break-even point. Generally, in engineering projects investors consider a SPBP of 5 years as acceptable. The SPBP does not account for the time value of money; however it is a useful tool for the quick assessment of whether a project or system is a viable option.

$$\text{SPBP} = \frac{I_{cc}}{\text{Annual savings}} \tag{8.4}$$

I_{cc} is the is the initial capital cost of the system (£). Annual savings are calculated by subtracting the annual total cost (A_{TC}) of the base case system from the annual total cost of the proposed system.

Next, Sect. 8.2.2 presents the economic assessment results.

8.2.2 Economic Assessment Results

The following assumptions have been made for the economic assessment of the novel tri-generation and equivalent base case system.

- System lifetime (*N*): 15 years (CFCL 2009)
- SOFC CHP system cost and installation: £20,950 (Elmer et al. 2015b)
- SDCS cost: £2700
- Potassium formate solution cost (20 kg): £235
- SOFC stack replacement cost and system maintenance: £5000 every 5 years
- UK micro-CHP feed-in-tariff (FiT): 0.125 £ kWh^{-1} (DECC 2014)
- Boiler and installation cost: £1300
- VCS capital cost: £500 per kW of cooling (Infante Ferreira and Kim 2014)
- Annual VCS maintenance cost: 10 % of VCS capital cost
- Annual gas check: £60
- Average natural gas unit cost: 0.0421 £ kWh^{-1} (EST 2014)
- Average electricity unit cost: 0.172 £ kWh^{-1} (Goot 2013)
- Average yearly VCS COP_{el}: 2 (Welch 2008)
- Average heating system efficiency (boiler + distribution): 85.5 %
- Annual cooling time required: 1200 h $year^{-1}$ (Pilatowsky et al. 2011)
- Interest rate (i_r): 7 % (constant throughout assessment period)
- Inflation rate (i_f): 3 % (constant throughout assessment period)
- Scrap value (SV): 10 % of initial capital cost (Pilatowsky et al. 2011)

In the UK, fuel cell CHP of 2.0 kW_e or less qualifies for the micro-generation FiT (DECC 2014). Under this scheme, the UK government pays 0.125 £ kWh^{-1} of electricity generated, regardless of whether it is consumed or exported. Where relevant, the economic assessments consider the FiT.

Figure 8.1a, b shows the NPC of the 1.5 and 2.0 kW_e tri-generation systems and equivalent base case systems over a 15 year period. The assessment considers the performance of the tri-generation system with and without FiT support. The initial NPC in year 0 is the system investment cost, which is much higher for the tri-generation system compared to the base case. The NPC of the systems increases over time due to the annual operating costs. The tri-generation system with FiT support displays only a marginal increase in the NPC over the 15 year period because the FiT almost pays for the annual operating cost of the system. For the tri-generation systems, an NPC spike is seen at year five and ten; this is due to stack replacement. The small dip in NPC at year 15 is due to the scrap value of the systems.

Table 8.1 presents the NPC, EUAC and SPBP results for the tri-generation and base case systems.

Without FiT support, the NPC of both the 1.5 and 2.0 kW_e tri-generation system are 26 and 10 % higher than the equivalent base case system respectively. However, with FiT support there is a 31 and 90 % reduction in the NPC of the 1.5 and 2.0 kW_e tri-generation system compared to the equivalent base case system respectively. When the FiT is considered the annual revenue means the tri-generation system has a favourable NPC compared to the base case in year 11.5 for the 1.5 kW_e tri-generation system and year 7 for the 2.0 kW_e tri-generation system. The NPC of the 1.5 kW_e tri-generation system is lower than the 2.0 kW_e tri-generation system when no FiT is considered, but higher when the FiT is considered. The higher NPC seen in the 2.0 kW_e tri-generation system without FiT is

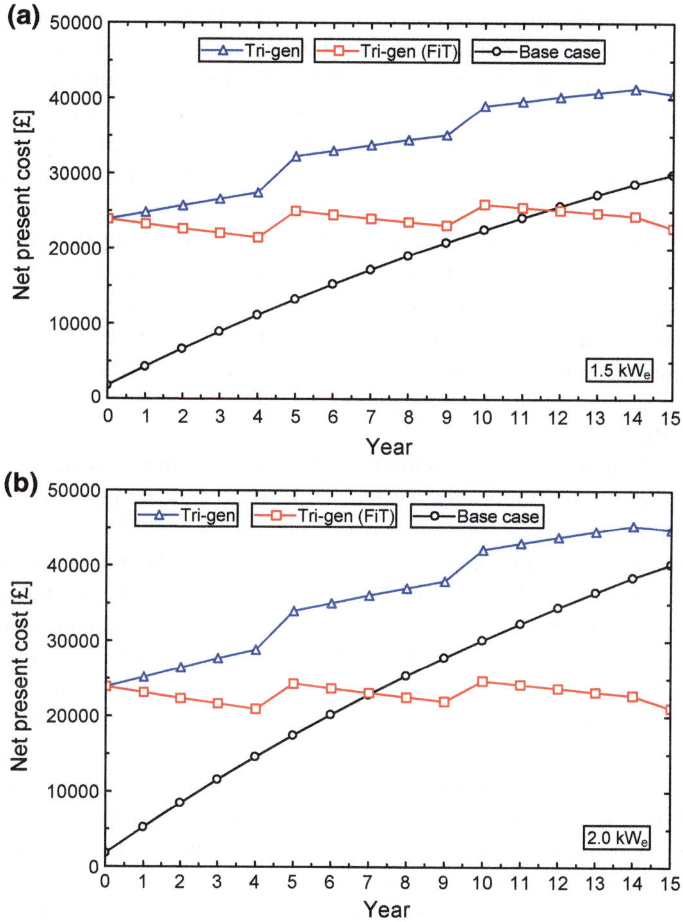

Fig. 8.1 NPC comparison at a 1.5 kW_e in **a** and 2.0 kW_e in **b** capacity between the tri-generation system with and without the FiT and the base case system

Table 8.1 Economic assessment results

	1.5 kW_e tri	1.5 kW_e base	2.0 kW_e tri	2.0 kW_e base
NPC (no FiT)	£40,544	£29,898	£44,818	£40,257
NPC (FiT)	£22,770	–	£21,120	–
EUAC (no FiT)	£4451	£3283	£4921	£4420
EUAC (FiT)	£2500	–	£2319	–
SPBP (no FiT)	19.8 years	–	14.7 years	–
SPBP (FiT)	9.8 years	–	7.3 years	–

due to the higher fuel input requirement, and thus higher annual operating costs. However, when FiT is considered the 2.0 kW_e tri-generation system provides greater annual revenues and thus a lower NPC. Both with and without FiT support, the 2.0 kW_e tri-generation system has a lower SPBP compared to the 1.5 kW_e tri-generation system. Although the 2.0 kW_e tri-generation system suffers an electrical efficiency reduction and thus a greater fuel input, the higher electrical capacity means it is offsetting more grid derived electricity. Per kWh, grid derived electricity has a higher associated cost compared to natural gas, and thus the SPBP of the 2.0 kW_e tri-generation system is lower. Furthermore, the 2.0 kW_e tri-generation system has a greater cooling output, and thus the equivalent base case system requires more grid derived electricity for the VCS. In all cases the tri-generation system generates annual operating cost savings compared to the base case system. The high NPC and SPBP of the tri-generation system are therefore due to the capital cost of the SOFC.

Figure 8.2a compares the economic performance of the 1.5 kW_e tri-generation system and equivalent base case system with respect to the unit cost of electricity. No FiT is considered. The unit cost of electricity does not affect the NPC of the tri-generation system, only the base case system. As the unit cost of electricity increases from 0.05 to 0.6 £ kWh^{-1} the NPC of the base case system increases, and thus the economic feasibility of the tri-generation system improves. At an electrical unit cost of 0.2458 £ kWh^{-1} there is a NPC break-even point between the tri-generation and base case system. Above 0.2458 £ kWh^{-1} the 1.5 kW_e tri-generation system has a better NPC and should be considered over the base case system. At an electrical unit cost of 0.2458 £ kWh^{-1} the tri-generation system has a SPBP of 12 years. For the SPBP to fall below 5 years, an electrical unit cost of 0.55 £ kWh^{-1} is required. In comparison, the 2.0 kW_e tri-generation system has a NPC break-even electrical unit cost of 0.1955 £ kWh^{-1}. Due to the continual rise in utility electricity prices, the break-even electrical unit costs which produce tri-generation system economic feasibility are realistic and not too far off current prices as demonstrated in Fig. 8.3.

Figure 8.2b compares the economic performance of the 1.5 kW_e tri-generation system and equivalent base case system with respect to the unit cost of natural gas. No FiT is considered. Natural gas unit cost affects both the tri-generation and base case systems NPC. As the unit cost of natural gas increases from 0.01 to 0.1 £ kWh^{-1} the NPC of both the tri-generation and base case systems increase. The tri-generation system is more sensitive to changes in the unit cost of natural gas compared to the base case system due to a greater proportionate demand. For the 1.5 kW_e tri-generation system there is not a natural gas unit cost that makes the tri-generation system favourable i.e. a NPC break-even point. As the natural gas unit price is increased the reduction in NPC between the base case and tri-generation system increases, and as a result the SPBP increases. As the natural gas unit cost is increased from 0.01 to 0.1 £ kWh^{-1} the tri-generation system SPBP increases from 14 to 51 years. The 2.0 kW_e tri-generation system does have a NPC break-even

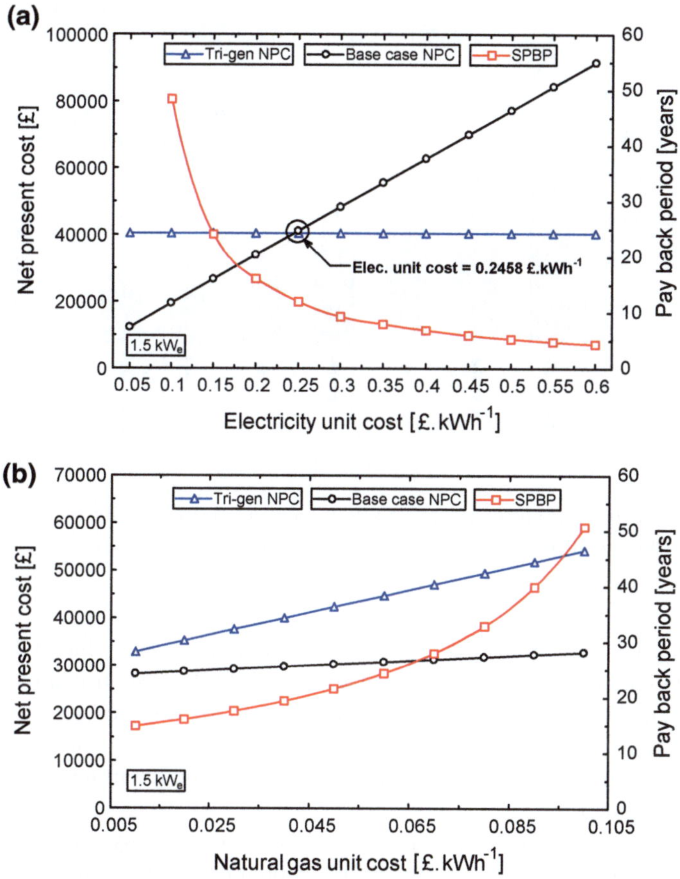

Fig. 8.2 NPC and SPBP comparison between the 1.5 kW_e tri-generation system and base case system with **a** electricity unit cost, and **b** natural gas unit cost

natural gas unit cost of 0.0233 £ kWh^{-1}. However this is very low and not realistic in the current economic climate where fossil fuels have such value.

Figure 8.3 shows the NPC of a 1.5 and 2.0 kW_e equivalent base case system in a range of different counties with respect to electrical unit cost data published by the International Energy Agency (IEA 2012). The NPC of the respective tri-generation systems (horizontal lines) are plotted to indicate which countries the novel system is currently economically viable in. Based on the current assumptions, the novel tri-generation system (1.5 and 2.0 kW_e) is only economically viable in Denmark where the unit cost of electricity is 0.262 £ kWh^{-1}. The largest different between the NPC of the tri-generation and base case system is in China, where the unit cost of electricity is as low as 0.0512 £ kWh^{-1}. Based purely on economic performance, the novel tri-generation system is more suited to European locations, where on average the unit cost of electricity is higher than Asia and the Americas.

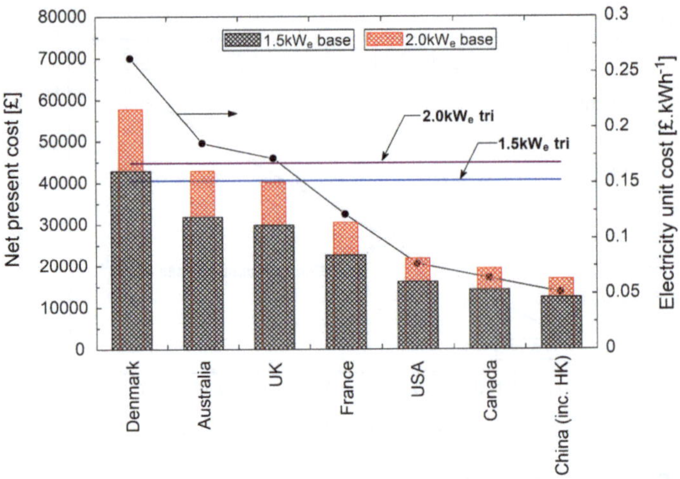

Fig. 8.3 NPC comparison between the 1.5 kW_e tri-generation system and base case system with respect to country of operation

As discussed in Fig. 8.2a, the 2.0 kW_e tri-generation system has a lower NPC break-even electrical unit cost. As a result, the 2.0 kW_e system is almost feasible in the current Australian economic climate. Section 8.3 assesses the environmental performance of the tri-generation in the same countries. The aim is to highlight any geographical differences between the economic and environmental feasibility of the novel system.

Figure 8.4a shows the NPC of the 1.5 kW_e tri-generation system and equivalent base case system with respect to the SOFC capital cost. The capital cost of the tri-generation system, operating at a 1.5 kW_e capacity, needs to be £9715 or less for it to be economically viable compared to the base case system. At a 2.0 kW_e capacity the required SOFC capital cost is £16135. As the capital cost of the SOFC increases, the SPBP increases. At the 1.5 kW_e NPC break-even point of £9715 the SPBP is 12.8 years. Although not shown in Fig. 8.4a, variation in the liquid desiccant system capital cost has a negligible impact on NPC and SPBP. Reducing the liquid desiccant system capital cost by 50 % results in a 4.5 % reduction in the SPBP. Reducing the SOFC capital cost by 50 % results in a 32 % reduction in the SPBP, demonstrating that tri-generation system economic viability presides with reducing the capital cost of the SOFC.

Figure 8.4b shows the NPC for the 1.5 kW_e tri-generation and equivalent base case system with respect to SOFC capital cost and unit cost of electricity respectively. Up to an electricity unit cost of 0.11 £ kWh^{-1} the base case system is always better than the tri-generation system. However at the electrical unit cost reference value of 0.172 £ kWh^{-1}, the 1.5 kW_e tri-generation system is competitive when the SOFC capital cost is less than £9500. At the intersection point, the tri-generation system is economically favourable if the SOFC capital

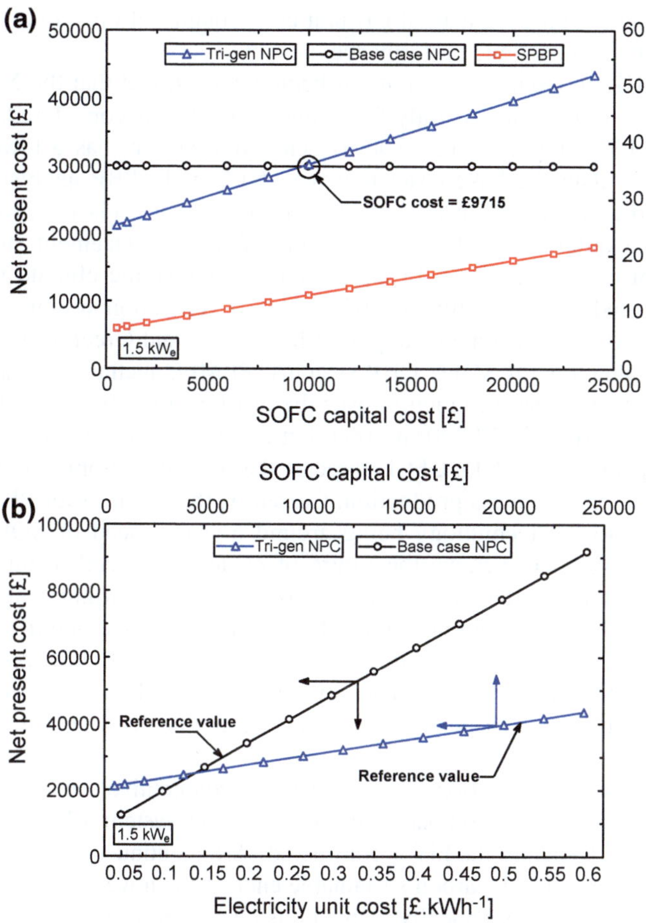

Fig. 8.4 NPC and SPBP comparison between the 1.5 kW_e tri-generation system and base case system with **a** SOFC capital cost, and **b** electricity unit cost and SOFC capital cost

cost is less than £4750 with an electrical unit cost of greater than 0.14 £ kWh^{-1} (i.e. UK, Australia).

Next, Sect. 8.2.3 contextualises the presented economic assessment results and concludes the section.

8.2.3 Economic Assessment Conclusions

Section 8.2 has presented an economic assessment comparing the novel tri-generation system to an equivalent base case system. NPC, EUAC and SPBP have been used as the means of assessment. Sensitivity analysis has been used to assess the impact electricity unit cost, natural gas unit cost, country of operation and SOFC capital cost has on economic performance.

Within a UK economic climate it has been demonstrated that the NPC of the novel tri-generation system is only favourable when FiT is considered, in which case the 2.0 kW_e output is best. The tri-generation system has a lower annual operating cost than the base case; however, NPC and SPBP analysis demonstrates that the novel system is currently uneconomical. This is primarily due to the SOFC capital cost and the requirement of stack replacement, not the liquid desiccant unit capital cost. In the current UK economic climate the SOFC capital cost needs to be less than £9000 for the tri-generation system to be competitive. This is a cost estimate supported by Staffell and Green (2013) in their economic evaluations of SOFC CHP systems. PEMFC technology has demonstrated considerable price reduction over the last 6 years. The 1 kW_e Panasonic unit had a unit cost of £27,300 in 2009, but as of 2015 it is being supplied to energy companies for £3600. CFCL forecast that they can supply the BlueGEN unit for £5200 once in mass production. Currently, the much lower PEMFC unit costs are due to the technology being around 5 years ahead of SOFC. Many commercial developers believe the future of cheaper fuel cell technology lies with SOFC systems as they do not need to use expensive platinum catalysts like PEMFC. Based on the example of PEMFC cost reductions, significant SOFC cost reductions can be anticipated. The SOFC cost target figures presented in this chapter are therefore sensible and could be realistically achieved in the next 5–10 years, making the tri-generation system economically viable in almost all cases.

Currently, the tri-generation system becomes competitive, and even demonstrates good profitability, compared to the base case system when government incubator support, such as the FiT is considered. With continued instability in governmental support for low carbon sustainable energy, the novel tri-generation system needs to become economically viable in its own right for it to be considered a viable alternative to conventional energy supply. Furthermore, a 2.0 kW_e base load capacity is large, and effective electrical utilisation may be problematic, particularly in a domestic building context. With the possibility of future withdrawal of government support for fuel cell CHP, maximising in-house electrical consumption will be essential to maintain economic viability. A lower electrical capacity fuel cell would therefore be required. The Japanese domestic market, which is estimated to be 10 years ahead of the European market, is now focussing domestic fuel cell CHP development at capacities of 750 W_e (Ellamla et al. 2015), a possible insight into the future of where European domestic fuel cell development needs to go.

Like other small scale tri-generation systems presented in the literature, the economic performance of the SOFC liquid desiccant tri-generation system is most sensitive to the unit cost of natural gas (Huangfu et al. 2007). The tri-generation system is economically superior, compared to the base case system, when the unit cost of electricity is greater than 0.24 £ kWh^{-1} and as a result Denmark is currently the only country investigated where the tri-generation is economically viable. However, with the extraction of easily accessible fossil fuels diminishing, the unit cost of electricity in many countries is set to continue to rise, thus strengthening the economic case of the tri-generation system (DECC 2013).

Next, Sect. 8.3 provides an environmental assessment of the tri-generation system.

8.3 Environmental Assessment

In this section, an environmental assessment of the novel tri-generation system operating within a UK energy system context is presented. The environmental assessment compares the 1.5 and 2.0 kW_e tri-generation system with an equivalent base case system comprising grid electricity, natural gas fired boiler and electrically driven VCS. The evaluation metric used in the environmental assessment is the annual CO_2 emission. This is determined through the multiplication of the annual natural gas and electrical demand by their respective emission factors and summing the result. The emission factors of natural gas and electricity are varied, in a reasonable range, to carry out a sensitivity analysis of the environmental performance. Using electrical emission factor data published by Brander et al. (2011), the environmental performance of the tri-generation system in the context of different countries is presented. The assumptions used in the environmental assessment are the same as those presented in the economic assessment in Sect. 8.2.2. The emission factors used are based on a UK energy system context, and are as follows:

- Average natural gas emission factor: 0.184 kg CO_2 kWh^{-1} (EST 2014)
- Average electricity emission factor: 0.555 kg CO_2 kWh^{-1} (AMEE 2014)

Table 8.2 presents the environmental assessment results. The 1.5 and 2.0 kW_e tri-generation systems produce a respective 51.3 and 50.2 % reduction in annual CO_2 emissions compared to the equivalent base case system.

Figure 8.5a shows the annual CO_2 emissions of the 1.5 and 2.0 kW_e tri-generation systems and equivalent base case systems with respect to natural gas emission factor. Over the investigated natural gas emission factor range of 0.05–0.3 kg CO_2 kWh^{-1}, the tri-generation system always has a lower annual CO_2 emission. Both the tri-generation and base case system have a natural gas requirement. However, the greater proportionate natural gas demand in the tri-generation system means its annual CO_2 emission reductions are more sensitive to changes

Table 8.2 Environmental assessment results

	Annual emissions (kg CO_2)
1.5 kW_e tri	4030
1.5 kW_e base	8282
2.0 kW_e tri	5756
2.0 kW_e base	11,567

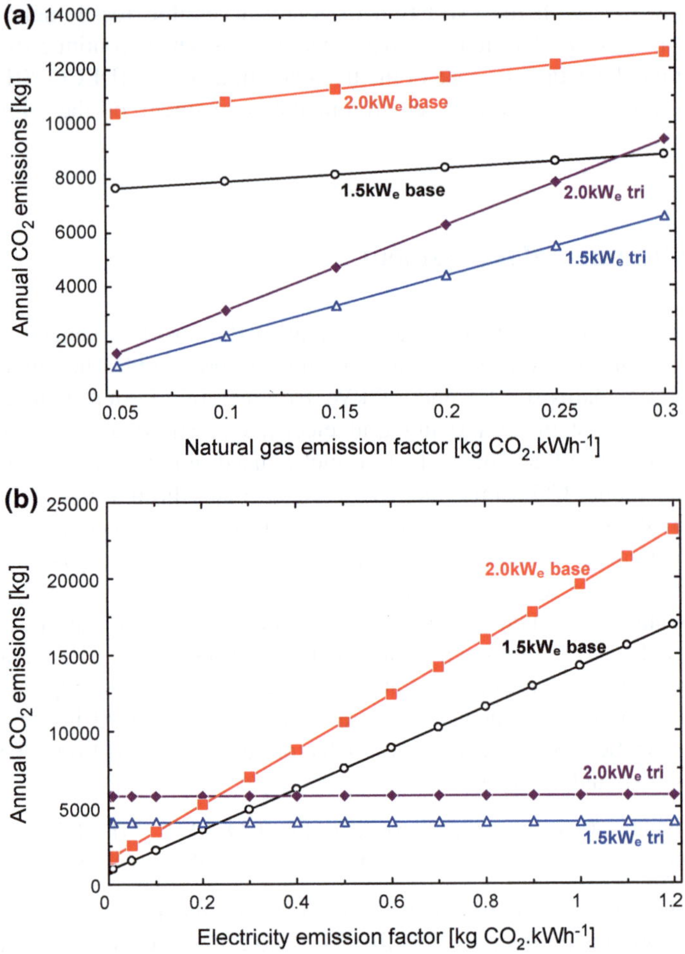

Fig. 8.5 Annual CO_2 emission comparison between the 1.5 and 2.0 kW_e tri-generation systems and equivalent base case system with **a** natural gas emission factor, and **b** electricity emission factor

in the natural gas emission factor. Consequently, as the natural gas emission factor is increased, the relative reduction in annual CO_2 emissions compared to the equivalent base case system is diminished. The 2.0 kW_e tri-generation system is more sensitive to changes in the natural gas emission factor than the 1.5 kW_e tri-generation system due to a lower electrical efficiency.

Figure 8.5b shows the annual CO_2 emissions of the 1.5 and 2.0 kW_e tri-generation systems and equivalent base case systems with respect to electrical emission factor. The tri-generation system has no electrical demand, and thus only the base case system is affected by the electrical emission factor. The tri-generation system has a lower annual CO_2 emission compared to the equivalent base case system when the electrical emission factor is greater than 0.2363 kg CO_2 kWh^{-1} for the 1.5 kW_e case and 0.2305 kg CO_2 kWh^{-1} for the 2.0 kW_e case.

Figure 8.6 shows the annual CO_2 emissions of the 1.5 and 2.0 kW_e equivalent base case system in a range of different counties using electrical emission factor data published by Brander et al. (2011). The annual CO_2 emissions of the respective tri-generation systems (horizontal lines) are plotted to indicate the countries in which the novel system is currently environmentally viable. The 1.5 and 2.0 kW_e tri-generation system is feasible in all the countries investigated except France and Norway as these countries have an average electrical emission factor of less than 0.1 kg CO_2 kWh^{-1}. France and Norway have an energy system that is largely characterised by the use of nuclear and renewables. As a result, the average electrical emission factor is low. Figure 8.6 shows that the 1.5 and 2.0 kW_e tri-generation system is most environmentally viable in Australia and China. Australia and China generate a large proportion of their electricity from coal, which has a high emission factor per kWh of electricity generated, and thus strengthens the environmental benefit of adopting the novel tri-generation system. Based on the data presented in Figs. 8.3 and 8.6, Denmark is currently the only country investigated where the novel tri-generation system is both economically and environmentally viable. Interestingly, the countries where the tri-generation system is not economically feasible due to a low electrical unit cost are in general the countries in which the system is most environmentally feasible i.e. Australia and China. This is primarily due to cheap electrical generation from easily accessible, more polluting fuels such as low grade coal.

Next, Sect. 8.3.1 contextualises the presented environmental assessment results and concludes the section.

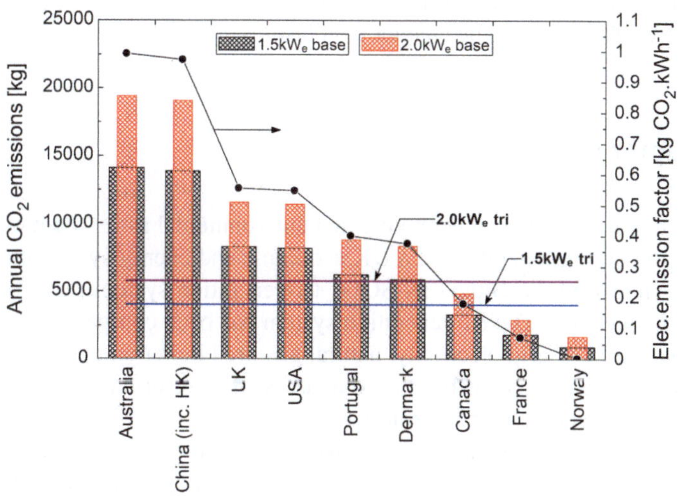

Fig. 8.6 Annual CO_2 emission comparison between the 1.5 and 2.0 kW_e tri-generation systems and equivalent base case system with respect to country of operation

8.3.1 *Environmental Assessment Conclusions*

Section 8.3 has presented an environmental assessment, based on annual CO_2 emissions, of the novel tri-generation system compared to an equivalent base case system. Sensitivity analysis has been used to assess the impact natural gas emission factor, electricity emission factor and country of operation has on environmental performance.

The environmental assessment has demonstrated that the tri-generation system is environmentally viable in almost all scenarios. In a UK energy system context the tri-generation system generates up to 51 % annual CO_2 emission reductions compared to the base case. Over the investigated natural gas emission factor range, the tri-generation system is always superior. The tri-generation system's environmental performance is not directly influenced by changes in the electrical emission factor, however the base case is. As a result, changes in the electrical emission factor have a marked impact on the relative performance of the tri-generation system with respect to the base case system. The tri-generation system is environmentally viable when the electricity emission factor is greater than 0.23 kg CO_2 kWh^{-1}. France and Norway have a large nuclear and renewable (hydro-electric) energy capacity. As a result, their electricity emission factor is low, and thus the tri-generation system does not provide an environmental benefit in such a setting. Countries such as Australia and China demonstrate the greatest environmental benefit from adopting the novel tri-generation system. However, as Berger (2015) states, the move to a hydrogen economy and with it the transition from the use of hydrocarbon to pure hydrogen-fed fuel cells in the next 30 years provides the potential for highly efficient, zero carbon energy conversion. With such a transition the novel tri-generation system would be highly competitive in almost all scenarios. This is discussed further in Sect. 9.4.

Next, Sect. 8.4 presents the chapter conclusions.

8.4 Chapter Conclusions

This chapter has provided an economic and environmental assessment of the novel tri-generation system operating in a UK economic and energy system context. The assessment has used the BlueGEN SOFC tri-generation system performance data presented in Sect. 7.2. The tri-generation system has been compared to a base case system comprised of grid electricity, natural gas fired boiler and electrically driven VCS. Sensitivity analysis has been used to assess the performance of the tri-generation system across a range of operating scenarios.

The economic assessment has demonstrated that the novel tri-generation system is viable only in certain cases. The tri-generation system has a lower annual operating cost than the base case, however, the high capital cost of the SOFC and requirement of stack replacement means that the tri-generation system NPC is only favourable when FiT is considered. However, with anticipated SOFC capital cost reductions the economic performance is predicted to improve. The current tri-generation system does not have a SPBP of less than 5 years, and is thus not immediately attractive to investors. Furthermore, with the possibility of future withdrawal of government support, a move mirroring the Japanese market towards smaller electrical capacity domestic fuel cells may be required to achieve/maintain economic viability. The economic performance of the tri-generation system is sensitive to natural gas and electrical unit cost. The future economic feasibility of the system will therefore be dependent upon future energy prices, which can be highly volatile. Currently, the tri-generation system is only economically viable in Denmark due its high unit cost of electricity.

The environmental assessment has demonstrated that the novel tri-generation system is viable across a large range of operational values. Within a UK energy system context, annual CO_2 emission reductions of up to 51 % compared to the equivalent base case system have been demonstrated. The environmental performance of the tri-generation system is more sensitive to changes in the natural gas emission factor than the base case system. The CO_2 emissions of the tri-generation system are insensitive with respect to electricity emission factor. However, electricity emission factor does affect the relative performance of the tri-generation system with respect to the equivalent base case system. The tri-generation system is environmental superior when the electricity emission factor is greater than 0.23 kg CO_2 kWh^{-1}. As a result, the tri-generation system is not currently viable in France and Norway. Australia and China demonstrate the greatest environmental benefit from adopting the novel tri-generation system. With a transition to hydrogen-fed fuel cells, the novel tri-generation system will be highly competitive in almost all scenarios.

This chapter has provided a detailed economic and environmental assessment of the novel tri-generation system. The following general conclusions, with respect to the chapter aims set out in Sect. 8.1, are as follows:

(1) The system is currently only economically viable with government support. SOFC capital cost and stack replacement are the largest inhibitors to economic viability. Environmental performance is closely linked to electrical emission factor, and thus performance is heavily country dependent.
(2) The countries, in which the system is environmentally viable, are in general the counties in which the system is not economically feasible. This is primarily due to the play off between cheap electrical generation from fossil fuels and more expensive cleaner electrical generation from renewables or nuclear.
(3) The economic and environmental feasibility of the novel tri-generation system will improve with predicted SOFC capital cost reductions and the transition to clean hydrogen production.

Although the SDCS has been developed with the aim of integration alongside a SOFC into a complete tri-generation system, the SDCS shows significant potential for integration with other CHP prime mover technologies such as ICE or SE. Due to the lower capital cost of ICE and SE technology (roughly ten times that of SOFC) and cheaper maintenance/part replacements, the economic performance of an ICE/SE based liquid desiccant tri-generation system can be expected to be much better than the current SOFC based system. However, the environmental performance of the SOFC based system will remain favourable compared to alternative options due to high electrical conversion efficiency and the provision for zero carbon energy conversion with the transition to a pure hydrogen fuel feed.

This chapter has demonstrated that the novel tri-generation system is, in certain cases, economically and environmentally viable. Next, Chap. 9 presents the conclusions to the thesis.

References

AMEE. 2014. *Realtime carbon*. Retrieved 06/04/2014, 2014, from http://realtimecarbon.org/.

Berger, R. 2015. *Advancing Europe's energy systems: Stationary fuel cells in distributed generation. F. C. A. H. J. Undertaking*. Luxembourg: Publications Office of the European Union. 2015.

Brander, M., A. Sood, C. Wylie, A. Haughton and J. Lovell. 2011. Technical Paper|Electricity-specific emission factors for grid electricity. *Ecometrica*.

CFCL. 2009. *Electronic resource: CFCL BlueGEN modular generator—Power and heat*. Electronic resource, from http://www.bluegen.info/Assets/Files/BlueGen_Brochure_(ENG_GER)_April_2010.pdf. 12 Aug 2012

DECC. 2013. *Estimated impacts of energy and climate change policies on energy prices and bills. D. O. E. A. C. Change*. London: Crown Copyright.

DECC. 2014. *CHP Incentives—Feed-in Tariff*. Retrieved 06/04/2014, 2014, from http://chp.decc.gov.uk/cms/feed-in-tariff/.

EST. 2014. *Energy saving trust—Our calculations*. From http://www.energysavingtrust.org.uk/Energy-Saving-Trust/Our-calculations.

Ellamla, H.R., I. Staffell, P. Bujlo, B.G. Pollet, and S. Pasupathi. 2015. Current status of fuel cell based combined heat and power systems for residential sector. *Journal of Power Sources* 293: 312–328.

Elmer, T., M. Worall, S. Wu, and S.B. Riffat. 2015. Fuel cell technology for domestic built environment applications: State of-the-art review. *Renewable and Sustainable Energy Reviews* 42: 913–931.

Elmer, T., M. Worall, S. Wu, and S. Riffat. 2016. Assessment of a novel solid oxide fuel cell tri-generation system for building applications. *Energy Conversion and Management* 124: 29–41.

Goot, G. 2013. *Electricity Price per kWh* (2013). Retrieved 06/03/2014, 2014, from http://blog.comparemysolar.co.uk/electricity-price-per-kwh-2013-comparison-of-e-on-edf-npower-british-gas-scottish-and-sse/.

Huangfu, Y., J.Y. Wu, R.Z. Wang, X.Q. Kong, and B.H. Wei. 2007. Evaluation and analysis of novel micro-scale combined cooling, heating and power (MCCHP) system. *Energy Conversion and Management* 48(5): 1703–1709.

IEA. 2012. *Electricity information*. IEA Statistics.

Infante Ferreira, C., and D.-S. Kim. 2014. Techno-economic review of solar cooling technologies based on location-specific data. *International Journal of Refrigeration* 39: 23–37.

Pilatowsky, I., R.J. Romero, C.A. Isaza, S.A. Gamboa, P.J. Sebastian, and W. Rivera. 2011. *Cogeneration fuel cell—sorption air conditioning systems*. London: Springer.

Staffell, I., and R. Green. 2013. The cost of domestic fuel cell micro-CHP systems. *International Journal of Hydrogen Energy* 38(2): 1088–1102.

Welch, T. 2008. *CIBSE Knowledge Series: KS13—Refrigeration*. ed. H. Carwarardine and K. Butcher. London: CIBSE Publications.

Chapter 9
Conclusions and Future Work

9.1 Introduction

In response to the critical need to decarbonise the built environment, alternative methods for more effective energy utilisation have been explored using highly efficient decentralised tri-generation systems. Compared to a conventional separate generation scenario, energy conversion in a tri-generation system offers the potential for high energy efficiency and thus greater energy utilisation, providing a range of technical, environmental and economic benefits.

The aim of this thesis has been to design, develop and test an efficient and effective proof of concept tri-generation system based on SOFC and liquid desiccant air conditioning technology for building applications. The integration of SOFC and liquid desiccant into the first of its kind tri-generation system provides a clear contribution to new knowledge in the field of low carbon sustainable energy conversion techniques for building applications. In summary, the thesis has demonstrated the tri-generation system concept, and proved that SOFC and liquid desiccant technology are a viable technological combination.

In order to bring the thesis to a successful and coherent conclusion this chapter is split into three sections. Section 9.2 revisits the thesis aim and objectives, and critically examines the degree to which they have been fulfilled. Section 9.3 presents the thesis recommendations for future work. Finally, Sect. 9.4 concludes the thesis by taking a step back, in order to review the potential contribution of this tri-generation system in a wider context.

T. Elmer, *A Novel SOFC Tri-generation System for Building Applications*,
Springer Theses, DOI 10.1007/978-3-319-46966-9_9

9.2 The Degree of Completion of the Thesis Aim and Objectives

The preceding eight chapters have sought to achieve the thesis aim through the accomplishment of the seven thesis objectives as set out in Sect. 1.5. In this section the seven thesis objectives and the degree to which they have been accomplished are discussed, with particular reference to contribution to new knowledge. Following this, the achievement of the thesis aim is critically examined.

9.2.1 Degree of Achievement of the Thesis Objectives

Objective one of highlighting the current gap in the literature surrounding SOFC liquid desiccant tri-generation systems, particularly for building applications, has been achieved through an extensive and rigorous review presented in Chap. 2. The review has established that no previous work relating to a SOFC liquid desiccant tri-generation system has been reported.

Objective two of selecting an appropriate working fluid for the liquid desiccant air conditioning system has been achieved through the evaluation of lithium chloride, calcium chloride and potassium formate in Chap. 3. The three desiccant solutions have been evaluated with respect to their potential for application in the novel tri-generation system. Potassium formate, operating at a 0.65–0.7 solution mass concentration has been identified as a suitable desiccant solution due to its good dehumidification capacity, low temperature regeneration requirement, negligible environmental impact, low corrosiveness and cost.

Objective three of numerical evaluation of the liquid desiccant air conditioning system, with particular regards to its suitability for integration in a SOFC tri-generation system has been achieved in Chap. 3. The modelling has demonstrated at cooling capacities of less than 10 kW (a) excellent cooling and humidity control, and (b) effective use of low grade thermal energy with a low temperature regeneration requirement of less than 60 °C. It has been highlighted that a clear operational advantage of the novel SOFC liquid desiccant tri-generation system is the potential for nonsynchronous operation. Re-concentrating the desiccant solution is an effective means of storing the constant thermal output from the SOFC, and has the potential to bring about improvements to system performance. Whilst remaining within the boundaries of realistic operating values, COP_{th} of 0.8 are attainable at a regenerator solution temperature of 40–50 °C, demonstrating the potential for effective SOFC tri-generation system integration.

Objective four of validating, theoretically, the feasibility of combining SOFC and liquid desiccant air conditioning technology into an efficient and effective tri-generation system has been accomplished in Chap. 4. It has been demonstrated that whilst working in a SOFC tri-generation system context, a potassium formate solution mass concentration of 0.65 is appropriate to facilitate balanced desiccant system operation. Based on realistic operating values, the novel 1.5 kW_e tri-generation system can achieve an overall efficiency of 78.98 % in CHP mode and 70.07 % in tri-generation cooling mode. These are values competitive with other micro tri-generation systems presented in the literature. The inclusion of liquid desiccant air conditioning provides an efficiency increase of up to 24 % compared to SOFC electrical operation only. The theoretical tri-generation system analysis has demonstrated that compared to the SOFC CHP component there are more operational variables that may be controlled in the liquid desiccant air conditioning component. As a result, it is the desiccant system's operation that should be optimised for successful tri-generation system integration.

The novel tri-generation system is comprised of two main components; the SOFC and liquid desiccant. Objective five of evaluating these two components using experimental data has been achieved. Chapter 5 presents a novel IDCS (integrated liquid desiccant air conditioning system). The aim of the IDCS was to overcome the often cited barriers of liquid desiccant application in tri-generation and building applications, and thus facilitate effective tri-generation system development. The IDCS demonstrates good dehumidification/cooling output, however issues of mass imbalance in the regenerator, desiccant solution leakage and poor controllability of operating variables meant the unit was deemed unsuitable for tri-generation system integration. As a result of these shortcomings a SDCS (separate liquid desiccant air conditioning system) was acquired and tested. Chapter 6 presents a detailed SDCS component evaluation. Results demonstrate good dehumidification ability, mass balance between the dehumidifier and regenerator, no desiccant solution leakage and a good control of operating variables. Testing of the SDCS within an environmental chamber simulates real life operation in a hot and humid climate. Using regenerator thermal input values typical of a SOFC CHP system, COP_{th} values of up to 0.66 have been reported, demonstrating the potential for effective tri-generation system integration.

Fulfilment of the objective of SOFC component evaluation has been achieved and is presented in Chap. 7. The original, building-installed 1.5 kW_e BlueGEN SOFC had a variety of technical problems, and as a result of commercial issues the SOFC could not be repaired. Electrical field trial data of the 1.5 kW_e SOFC over an 8 months period shows stable operation during this period with electrical efficiency of 55–60 %, and availability for power generation of 91.7 %. Long term stable operation and a good thermal agreement between the 1.5 kW_e SOFC output and required SDCS input indicate the potential for efficient and effective tri-generation system development. In response to the 1.5 kW_e BlueGEN SOFC failure, a 250 W_e micro-tubular SOFC had to be acquired. Following issues of sulphur poisoning, experimental testing results show that the micro-tubular SOFC has an electrical output of 150 W, at a net electrical efficiency of 11.65 %. Operating with a 2 L min^{-1} water volumetric flow in the WHR circuit, flow temperatures of up to 65 °C are possible, with a thermal output of 450W recorded. The thermal output is low, but is sufficient to demonstrate the tri-generation system concept.

Chapter 7 has successfully achieved objective six of integrating SOFC and liquid desiccant air conditioning technology into a complete tri-generation system.

The developed system has been evaluated on an energetic, economic and environmental basis. The 1.5 kW_e BlueGEN SOFC was not available for tri-generation system development. However, Sect. 7.2 presents a theoretical integration analysis based on collected empirical 1.5 kW_e BlueGEN SOFC and SDCS data. At a 1.5 kW_e output, a CHP efficiency of 81.6 % and a tri-generation efficiency of 68.9 % have been demonstrated, values competitive with other systems of this capacity reported in the literature and tri-generation system simulations presented in Chap. 4. The CHP and tri-generation system efficiency increases to 84 and 71.1 % respectively at a 2.0 kW_e output, however, the electrical efficiency decreases from 60 to 56 %, resulting in lower cost and emission savings. The inclusion of liquid desiccant air conditioning provides an efficiency increase of 9–15 % compared to SOFC electrical operation only, demonstrating the merit of the novel tri-generation system in applications that require simultaneous electrical power, heating and dehumidification/cooling. The novel tri-generation system concept is demonstrated experimentally using the micro-tubular SOFC. The experimental results demonstrate regeneration of the potassium formate solution using the thermal output from the micro-tubular SOFC in the first of its kind tri-generation system. The low thermal output from the micro-tubular SOFC means the moisture addition rate is low (0.11 g s^{-1}). However, balanced operation with the dehumidifier is possible. The novel system can generate 150.4 W of electrical power, 442.6 W of heat output or 278.6 W of cooling. Instantaneous tri-generation system efficiency is low at around 25 %. This is primarily due to the low capacity and poor performance of the micro-tubular SOFC. The SDCS COP_{th} is 0.62, an encouraging value for a waste heat driven cooling system of this capacity. The inclusion of liquid desiccant air conditioning provides an efficiency increase of up to 13 % compared to SOFC electrical operation only.

The thesis has established that a clear operational advantage of the novel SOFC liquid desiccant tri-generation system is the potential for nonsynchronous operation. Using this concept the experimental system can generate an increased peak cooling output of up to 527 W and a daily tri-generation efficiency of 37.9 %. This is an encouraging value for a tri-generation system of this capacity and serves to demonstrate the novel tri-generation system operating in a building application. The micro-tubular SOFC was acquired at short notice to replace the 1.5 kW_e BlueGEN SOFC. As seen in the low thermal output, it is not the ideal match for the developed SDCS. However, the novel tri-generation concept has been successfully demonstrated. SOFC and liquid desiccant air conditioning technology are a feasible pairing. Future work should look to improved optimisation and pairing of the components.

Chapter 8 presents a detailed economic and environmental assessment, comparing the 1.5 kW_e BlueGEN tri-generation system and a base case system. The tri-generation system's annual operating cost is significantly lower than the base case. However, NPC and SPBP analysis demonstrates that the novel system is currently uneconomical in a UK economic climate. This is primarily due to the high capital cost of the SOFC and the requirement of stack replacement. An SOFC capital cost of £9000 or less is required for the tri-generation system to be feasible.

Government incubator support makes the tri-generation system economically viable. Use of the SDCS with ICE or SE technology would provide much better economic performance due to significantly lower system capital cost. The tri-generation system demonstrates good environmental performance. Emission reductions of up to 51 % compared to the base case system have been presented. The encouraging emission reductions are primarily due to the high electrical efficiency of the SOFC and the replacement of electrically derived cooling with waste heat driven cooling.

Objective seven of recommending future work related to the novel tri-generation system is presented in Sect. 9.3. However, the technical and commercial issues encountered with the building-installed SOFC in Sect. 7.2 highlight the real challenge of fuel cell deployment in the built environment. Reliability, durability and cost currently pose a great barrier to the wider use of fuel cell technology and demonstrate the need to focus future work on addressing these issues.

9.2.2 Degree of Achievement of the Thesis Aim

Through accomplishment of the seven thesis objectives, the aim of the design, development and testing of an efficient and effective SOFC liquid desiccant tri-generation system has been successfully achieved. The novel system is truly multi-functional, designed to provide the energy needs of a building throughout all seasons. The system increases the utilisation of thermal energy from the SOFC during periods of low/no thermal demand, to provide a comfortable indoor building environment through dehumidification/cooling. Theoretical and experimental methods have been adopted in order to carry out a rigorous and analytical piece of work. The thesis has proved, both theoretically and experimentally, the novel tri-generation system concept. The relationship between SOFC and liquid desiccant component operation and tri-generation system performance has been documented. The climatic, economic and environmental conditions in which the system performs best are provided, and recommendations for future work in order to improve system performance have been discussed.

The thesis has been approached in a systematic and logical manner, with research integrity demonstrated throughout. As highlighted in the approach taken to the review of the literature presented in Chap. 2, the novel nature of the developed tri-generation system means there is no previous work directly related to it. It is therefore proposed that the thesis provides a clear contribution to new knowledge, and provides a step forward in the area of low carbon sustainable energy conversion techniques for building applications. Although there is no work directly relating to the thesis topic, it is suggested that the presented work builds upon the following fields of research; small (micro) scale tri-generation systems, particularly those adopting fuel cell (SOFC) technology and the application of liquid desiccant air conditioning in small (micro) scale tri-generation systems (Huangfu et al. 2007; Pilatowsky et al. 2011; Jradi and Riffat 2014c; Wu et al. 2014; Buker et al. 2015).

To summarise Sect. 9.2, the thesis presents the following general conclusions with respect to the novel SOFC liquid desiccant tri-generation system:

- SOFC and liquid desiccant air conditioning are an effective technological pairing. High tri-generation efficiencies, particularly in hot and humid climates, are demonstrated.
- Appropriate matching of component capacity is necessary. Overall tri-generation system performance is more influenced by the SOFC component than the liquid desiccant.
- It is primarily the optimisation of the liquid desiccant component that facilitates effective tri-generation system integration and operation. The inclusion of liquid desiccant can bring significant improvement to system performance.
- The developed SDCS shows significant potential for integration with other CHP prime mover technologies (ICE, SE), with likely improvements to the reliability and economic performance.
- The novel tri-generation system can effectively provide electricity, heating or dehumidification/cooling from a single fuel source. Distribution losses are eliminated.
- Nonsynchronous operation is a particular advantage specific to a SOFC liquid desiccant tri-generation system. Thermal energy storage in the form of concentrated desiccant solution has the potential to bring about improvements to system performance.
- The concept has been proven at a small capacity but it is easily scalable.
- SOFC can internally reform natural gas, therefore reducing system complexity and demonstrating the simplicity of tri-generation system integration into current markets where natural gas distribution is widespread. Furthermore, the developed tri-generation system demonstrates great potential for decentralised locations where electrical grid connection is unavailable.
- Economic performance is questionable. This is anticipated to improve with SOFC capital cost reductions.
- Favourable environmental performance is demonstrated, with the potential for zero carbon operation with the transition to a pure hydrogen fuel.

Section 9.2 has re-visited the thesis aim and objectives and critically examines the degree to which they have been fulfilled. Next, Sect. 9.3 presents the thesis recommendations for future work.

9.3 Recommendations for Future Work

This section presents four recommendations for future work.

The experimental system demonstrates low tri-generation efficiency. This is primarily due to poor electrical performance from the micro-tubular SOFC, and the thermal capacity being too small to produce significant regeneration, and thus cooling capacity. As a result, future work should focus on achieving an improved (thermal) match between the SOFC and liquid desiccant components. Additionally, more extensive tri-generation system testing should be conducted to further investigate the relationship between SOFC and liquid desiccant component operation on overall performance.

As highlighted in Chap. 2, system integration, optimisation and energy utilisation is not a large issue in theoretical or laboratory based projects, thus high system efficiencies have been reported. However in a real working environment, effective energy utilisation can pose a serious challenge to system performance, specifically how to maximise the utilisation of system energy outputs. The experimental system has been tested and evaluated in a laboratory setting. Future work should aim to integrate the tri-generation system in a building context to demonstrate and quantify the benefits of the system in a real operational setting. Potential building settings include: domestic homes, schools or offices that require electrical power, heating and dehumidification/cooling.

The nonsynchronous operation of the tri-generation system has been highlighted as a significant advantage of a SOFC liquid desiccant tri-generation system. Future work should aim to investigate experimentally the nonsynchronous operating concept and thermal energy storage in the form of concentrated desiccant solution in a building context. Demonstration and quantification of the benefits of the nonsynchronous operation concept will serve to increase the feasibility of the novel SOFC liquid desiccant tri-generation system.

The novel IDCS presented in Chap. 5 demonstrates the potential of integrated liquid desiccant air conditioning systems. Future work should focus on using the system evaluation data to improve the design to facilitate improved mass balance between the dehumidifier and regenerator. Furthermore, better sealing techniques to prevent desiccant solution leakage should be investigated. Improvements to the IDCS concept could permit the wider use of liquid desiccant air conditioning in tri-generation and building applications.

Section 9.3 has presented the recommendations for future work. Finally, Sect. 9.4 concludes the thesis by taking a step back, and discussing the wider context into which the developed tri-generation system may operate and thus contribute to.

9.4 The Thesis in a Wider Context

The thesis has, quite necessarily, taken a focussed and technical approach to the development of the novel tri-generation system. However, this section takes a step back and critically evaluates the presented work in a wider context.

The hydrogen economy was discussed in Sect. 1.4. The developed tri-generation system has a clear role to play in a hydrogen economy. Currently, the majority of fuel cell applications operate on hydrogen derived from steam reforming of a hydrocarbon fuel or, in the case of a SOFC, internally reforming a hydrocarbon fuel. This is the cheapest form of production; however it has the disadvantage of emitting modest amounts of environmentally harmful pollutants. Many countries employing large scale renewable energy devices face the issue of managing excess electrical capacity when supply is high and demand is low. A possible solution is to use excess renewable electrical capacity to generate clean hydrogen through the electrolysis of water. Hydrogen can be stored inter-seasonally to fill the gap between supply and demand. Due to increased renewable energy capacity, the falling unit cost of renewable electricity and improved electrolysis technology; the large scale transition to clean, zero carbon hydrogen production using the electrolysis of water is expected in the next 30 years (Berger 2015). It is proposed that the current use of hydrocarbon-fed fuel cells is an essential stepping stone in developing and refining the necessary technology for when the wide-spread transition to clean zero carbon hydrogen production can occur.

This thesis has presented the design, development and testing of a novel SOFC liquid desiccant tri-generation system that can provide highly efficient, zero carbon energy conversion in a decentralised manner. The developed system is a viable option for applications that require simultaneous electrical power, heating and dehumidification/cooling. Thus this thesis demonstrates the potential role of the tri-generation system in the transition to a future hydrogen economy.

Over forty years ago Schumacher in his seminal book *Small is Beautiful* (Schumacher 1974) argues the vision for a transition towards more efficient energy utilisation and away from a dependence on energy derived from fossil fuel as a global necessity. He puts forward the point that humanity is treating fossil fuels as income items even though they are capital. Furthermore, Schumacher states that one of the most fateful errors of our age is the belief that "*the problem of production*" has been solved. We need to consider not the fossil fuel resource as income but renewable energy resources as income.

This thesis commenced with an assertion based on this vision:

> Humanity is now at a global tipping point. Serious decisions regarding the future of world energy supply and demand need to be made.

The work presented in this thesis provides one small and very specific way to contribute to this vision; but a contribution nonetheless.

References

Berger, R. 2015. *Advancing Europe's energy systems: Stationary fuel cells in distributed generation. F. C. a. H. J. Undertaking*. Luxembourg: Publications Office of the European Union. 2015.

Buker, M.S., B. Mempouo, and S.B. Riffat. 2015. Experimental investigation of a building integrated photovoltaic/thermal roof collector combined with a liquid desiccant enhanced indirect evaporative cooling system. *Energy Conversion and Management* 101: 239–254.

Huangfu, Y., J.Y. Wu, R.Z. Wang, X.Q. Kong, and B.H. Wei. 2007. Evaluation and analysis of novel micro-scale combined cooling, heating and power (MCCHP) system. *Energy Conversion and Management* 48(5): 1703–1709.

Jradi, M., and S. Riffat. 2014c. Experimental investigation of a biomass-fuelled micro-scale tri-generation system with an organic Rankine cycle and liquid desiccant cooling unit. *Energy* 71: 80–93.

Pilatowsky, I., R.J. Romero, C.A. Isaza, S.A. Gamboa, P.J. Sebastian, and W. Rivera. 2011. *Cogeneration fuel cel—Sorption air conditioning systems*. London: Springer.

Schumacher, E.F. 1974. *Small is beautiful, a study of economics as if people mattered.* London: Sphere Books Ltd.

Wu, J.Y., J.L. Wang, S. Li, and R.Z. Wang. 2014. Experimental and simulative investigation of a micro-CCHP (micro combined cooling, heating and power) system with thermal management controller. *Energy* 68: 444–453.

Appendix A
Potassium Formate Property Calculations

A.1 Potassium Formate Vapour Pressure

This section describes the routine used to calculate the vapour pressure of the potassium formate solution (p_{sol}), with respect to solution mass concentration (X_{sol}) and temperature (T_{sol}). The routine is taken from the work of James (1998), and is based upon theoretical and empirical data.

The water present in the desiccant solution is calculated using Eq. A.1.

$$X_{H_2O} = 100 - X' \tag{A.1}$$

X' is the desiccant solution mass concentration as a percentage. The molecular weight of the desiccant solution, X_m, is calculated using Eq. A.2.

$$X_m = \frac{1}{1 + \left(\frac{M_{H_2O} \times X'}{M_{CHKO_2} \times X_{H_2O}}\right)} \tag{A.2}$$

M_{H_2O} and M_{CHKO_2} are the molecular weight of water vapour (18.015 g mol^{-1}) and the potassium formate solution respectively (84.11 g mol^{-1}).
A, B and Z are constants, calculated using Eqs. A.3, A.4 and A.5 respectively.

$$B = -3.42 - \left(\frac{1039}{T_{sol,K}}\right) \tag{A.3}$$

$$A = -2.08 - \left(\frac{1336}{T_{sol,K}}\right) \tag{A.4}$$

$$Z = \frac{A \times X_m}{A \times X_m + B \times (1 - X_m)} \tag{A.5}$$

T. Elmer, *A Novel SOFC Tri-generation System for Building Applications*,
Springer Theses, DOI 10.1007/978-3-319-46966-9

$T_{\mathrm{sol,K}}$ is the solution temperature in °K. γ is the activity coefficient of water and is calculated using Eq. A.6.

$$\gamma = A \times (1 - Z)^2 \times [1 + 4.19 \times Z \times (Z - 2/3)] \quad \text{(A.6)}$$

P_{H20} is the vapour pressure (Pa) of the water in the desiccant solution at a given temperature, and is calculated using Eq. A.7.

$$p_{\mathrm{H_2O}} = 73.649 - \left(\frac{7258.2}{T_{sol,K}}\right) - 7.3037 \cdot \ln\left(T_{sol,K}\right) + 0.0000041653\left(T_{sol,K}\right)^2 \quad \text{(A.7)}$$

p_{sol} is the vapour pressure (Pa) of the potassium formate desiccant solution, and is calculated using Eq. A.8.

$$p_{sol} = X_m \cdot \exp^{(\gamma)} \cdot \exp^{(p_{\mathrm{H_2O}})} \quad \text{(A.8)}$$

A.2 Potassium Formate Thermophysical Calculations

This section presents the linear regression curve functions fitted to the experimental data presented by Melinder (2007). The functions are used to calculate the thermophysical properties of the potassium formate desiccant solution, with respect to solution temperature (T_{sol}) and solution mass concentration (X_{sol}).

A.2.1 Potassium Formate Density

The density (kg m^{-3}) of the potassium formate solution is calculated using Eq. A.9.

$$\rho_{sol} = 1004 - 0.1191T_{sol} - 0.002{T_{sol}}^2 + 5.62X_{sol} + 0.02616{X_{sol}}^2 - 0.005975T_{sol} \times X_{sol} \quad \text{(A.9)}$$

A.2.2 Potassium Formate Specific Heat Capacity

The specific heat capacity (J kg^{-1} K) of the potassium formate solution is calculated using Eq. A.10.

$$c_{p,sol} = 4228.0298 - 1.485T_{sol} + 0.01122{T_{sol}}^2 + 0.0001042{T_{sol}}^3 - 44.49X_{sol} + 0.2287{X_{sol}}^2 + 0.0003806{X_{sol}}^3 + 0.1843T_{sol} \cdot X_{sol} - 0.002459T_{sol} \cdot {X_{sol}}^2 - 0.001094{T_{sol}}^2 \cdot X_{sol} + 0.00001727{T_{sol}}^2 \cdot {X_{sol}}^2 \quad \text{(A.10)}$$

A.2.3 Potassium Formate Thermal Conductivity

The thermal conductivity (W m^{-1} K) of the potassium formate solution is calculated using Eq. A.11.

$$
\begin{aligned}
k_{sol} = {} & 0.5646 + 0.001772 \cdot T_{sol} - 0.000006283 \cdot T_{sol}{}^2 + 1.669 \times 10^{-7} \cdot T_{sol}{}^3 - 2.778 \\
& \times 10^{-9} \cdot T_{sol}{}^4 - 0.001881 \cdot X_{sol} + 0.00003667 \cdot X_{sol}{}^2 - 0.000001146 \cdot X_{sol}{}^3 \\
& + 7.514 \times 10^{-9} \cdot X_{sol}{}^4 - 0.00002249 \cdot T_{sol} \cdot X_{sol} + 2.132 \times 10^{-7} \cdot T_{sol} \cdot X_{sol}^2 \\
& + 2.829 \times 10^{-10} \cdot T_{sol} \cdot X_{sol}{}^3 - 3.332 \times 10^{-7} \cdot T_{sol}{}^2 \cdot X_{sol} + 1.235 \\
& \times 10^{-8} \cdot T_{sol}{}^2 \cdot X_{sol}{}^2 - 1.159 \times 10^{-10} \cdot T_{sol}{}^2 \cdot X_{sol}{}^3 + 7.659 \\
& \times 10^{-9} \cdot T_{sol}{}^3 \cdot X_{sol} - 1.397 \times 10^{-10} \cdot T_{sol}{}^3 \cdot X_{sol}{}^2
\end{aligned}
\tag{A.11}
$$

A.2.4 Potassium Formate Dynamic Viscosity

The dynamic viscosity (mPa s) of the potassium formate solution is calculated using Eq. A.12.

$$
\begin{aligned}
\mu_{sol} = {} & 1.754 - 0.07023 \times 0.002794 \cdot T_{sol}{}^2 - 0.0001236 \cdot T_{sol}{}^3 + 0.000003162 \cdot T_{sol}{}^4 \\
& - 3.158 \times 10^{-8} \cdot T_{sol}{}^5 + 0.04564 \cdot X_{sol} - 0.004578 \cdot X_{sol}{}^2 + 0.0003114 \cdot X_{sol}{}^3 \\
& - 0.000008298 \cdot X_{sol}{}^4 + 8.078 \times 10^{-8} \cdot X_{sol}{}^5 + 0.0009856 \cdot T_{sol} \cdot X_{sol} \\
& - 0.00011305598 \cdot T_{sol} \cdot X_{sol}{}^2 + 0.000003868 \cdot T_{sol} \times X_{sol}{}^2 - 5.272 \\
& \times 10^{-8} \cdot T_{sol} \cdot X_{sol}{}^4 + 0.000006756 \cdot T_{sol}{}^2 \cdot X_{sol} - 4.128 \times 10^{-7} \cdot T_{sol}{}^2 \cdot X_{sol}{}^2 \\
& + 1.983 \times 10^{-8} \cdot T_{sol}{}^2 \cdot X_{sol}{}^3 + 5.727 \times 10^{-9} \cdot T_{sol}{}^3 \cdot X_{sol} - 1.188 \\
& \times 10^{-8} \cdot T_{sol}{}^3 \cdot X_{sol}{}^2 + 5.163 \times 10^{-9} \cdot T_{sol}{}^4 \cdot X_{sol}
\end{aligned}
\tag{A.12}
$$

A.3 Potassium Formate Henry's Law Constant

The Henry's Law constant (wt%) is a measure of the solubility of a gas in a liquid and is taken from experimental data presented in the literature (Stephen and Stephen 1963). The Henry's Law constant of the $CHKO_2$ solution is shown in Eq. A.13.

$$H_{CHKO_2} = 104.3 - 0.3828 T_{sol} \tag{A.13}$$

References

James, S. 1998. *A new working Fluid 'Potassium Formate' for use in absorption heat pumps.* Masters of Philosophy: The University of Nottingham.

Melinder, A. 2007. Thermophysical properties of aqueous solutions used as secondary working fluids. PhD Thesis, KTH Energy and Environmental Technology.

Stephen, H., and T. Stephen. 1963. Solubilities of inorganic and organic compounds, volume 1. *Binary Systems Part* 1(1): 179.

Appendix B
Thermodynamic Properties of Moist Air

In this section the equations used to calculate the thermodynamic properties of moist air are presented.

The air relative humidity, RH_a (%), is calculated using Eq. B.1.

$$RH_a = \frac{p_v}{p_{v,s}} \times 100\ \% \tag{B.1}$$

p_v and $p_{v,s}$ are the vapour pressure and saturation vapour pressure of the air respectively in Pa.

The air absolute humidity, ω_a (kg_{vapour}/kg_{dryair}), is calculated using Eq. B.2.

$$\omega_a = 0.622\frac{p_v}{p_a - p_v} \tag{B.2}$$

p_a is the pressure of the moist air in Pa (101,325 Pa)

The air specific enthalpy, h_a ($kJ\ kg^{-1}$), is calculated using Eq. B.3.

$$h_a = 1.005T_a + [\omega_a(2501 + 1.84T_a)] \tag{B.3}$$

T_a is the air temperature in °C

T. Elmer, *A Novel SOFC Tri-generation System for Building Applications*, Springer Theses, DOI 10.1007/978-3-319-46966-9

Appendix C
IDCS Inlet Test Conditions

The inlet air conditions for each of the IDCS tests are presented in Table C.1.

Table C.1 IDCS inlet air conditions

Test number	Dehumidifier		Regenerator	
–	T_a (°C)	RH_a (%)	T_a (°C)	RH_a (%)
1	30.73	50.04	24.694	55.182
2	30.92	50.59	26.035	51.819
3	30.17	51.54	26.003	46.228
4	30.54	51.52	26.521	46.975
5	30.58	60.02	24.289	49.238
6	30.71	60.3	23.53	44.02
7	30.67	60.19	22.998	42.273
8	30.17	70.5	24.751	39.972
9	30.29	70.09	25.148	38.361
10	30.32	70.29	25.096	37.874
11	35.43	47.31	24.563	41.137
12	36.24	50.22	23.899	38.433
13	36.33	51.77	25.126	49.387
14	34.54	60.48	22.35	51.109
15	35.99	60.45	25.233	43.632
16	35.47	59.56	25.192	44.784
17	35.33	70.41	24.533	50.498
18	35.02	68.73	23.434	66.301
19	34.71	70.37	24.119	62.559
20	27.71	63.03	23.822	48.248
21	30.94	61.85	24.592	48.358

T. Elmer, *A Novel SOFC Tri-generation System for Building Applications*,
Springer Theses, DOI 10.1007/978-3-319-46966-9

Appendix D
Micro-tubular SOFC Stability Testing Results

In this section the long term stability testing data of the micro-tubular SOFC is provided. The long term stability tests were carried out at the University of Birmingham (UoB). Figure D.1 provides a photograph of the test set up. During the stability testing the SOFC WHR was not tested.

Figure D.2a shows a photograph of the micro-tubular SOFC with labelled cathode gas inlet and gas outlet. WHR provision, presented in Chap. 7, passes this gas outlet through the recupertaor heat exchanger. Figure D.2b shows a photograph of the micro-tubular SOFC control and display panel. The switch button turns the SOFC on or off. The display panel shows the operational voltage of the SOFC, heat-up/cool down time, and the total number of unit operational hours.

Figure D.3 presents the long term stability testing data for the micro-tubular SOFC. Up until operational hour 75, the unit produces around 250 W of DC electrical power, at an electrical efficiency of 19.4 %. At hour 75 severe sulphur poisoning of the SOFC stack occurred resulting in no power output from the SOFC (no voltage). Once the sulphur traps were replaced, and the unit was operated for around 15 h, the SOFC recovers to a maximum power output of around 140 W DC. The micro-tubular SOFC data presented in Sect. 7.4.1 shows that the SOFC has now recovered a little more to a 150.4 W electrical output.

The long term stability testing results demonstrate the thermal resistance of the micro-tubular SOFC, and also its resilience to sulphur poisoning. Compared with planar-type SOFCs, these advantages show the potential of the future application of micro-tubular SOFCs.

T. Elmer, *A Novel SOFC Tri-generation System for Building Applications*,
Springer Theses, DOI 10.1007/978-3-319-46966-9

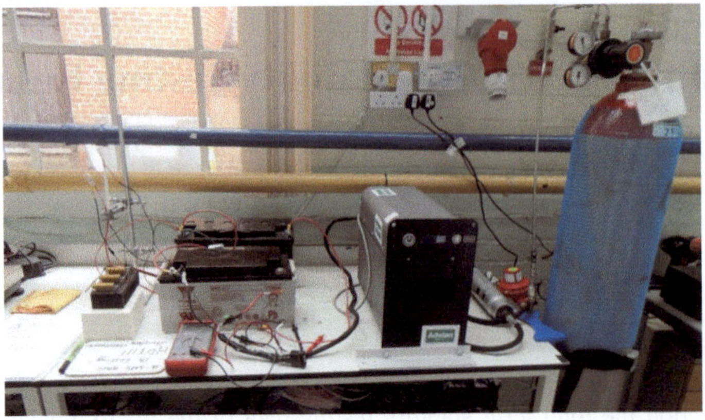

Fig. D.1 UoB micro-tubular SOFC test set-up

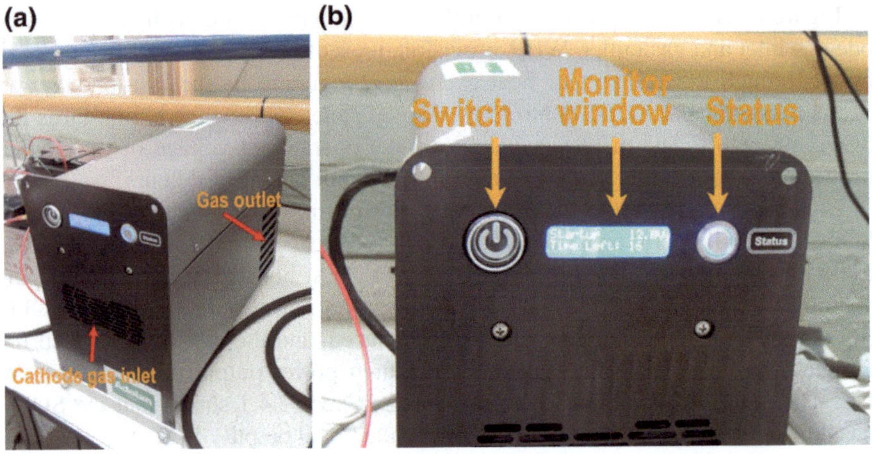

Fig. D.2 Micro-tubular SOFC, **a** gas inlet and outlet, and **b** control and display panel

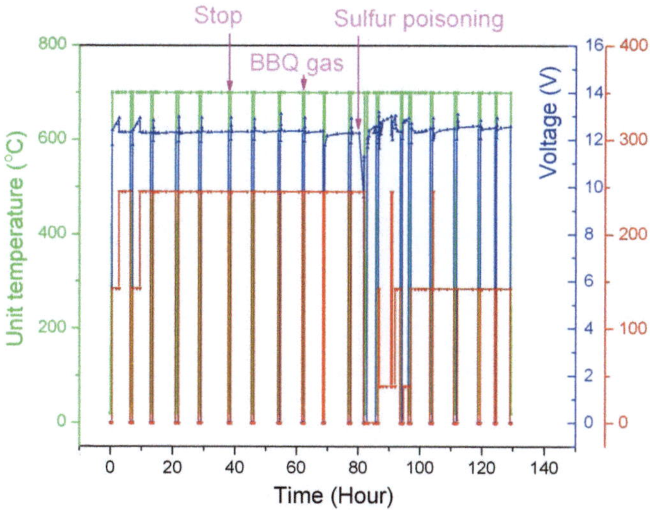

Fig. D.3 Micro-tubular SOFC long term stability test data

MIX
Papier aus verantwortungsvollen Quellen
Paper from responsible sources
FSC® C105338

If you have any concerns about our products,
you can contact us on
ProductSafety@springernature.com

In case Publisher is established outside the EU,
the EU authorized representative is:
Springer Nature Customer Service Center GmbH
Europaplatz 3, 69115 Heidelberg, Germany

Printed by Libri Plureos GmbH
in Hamburg, Germany